Digital Hostage

Frankie Shoultz

Published by Frankie Shoultz, 2026.

DIGITAL HOSTAGE

First edition. March 7, 2026.

ISBN: 979-8232572822

Written by Frankie Shoultz.

DIGITAL HOSTAGE

CHAPTER ONE

The New Surveillance Economy: *You Are the Product*

"If you are not paying for it, you are not the customer; you are the product being sold." Andrew Lewis, 2010

The Morning Harvest

It begins before your eyes open.

At 6:47 a.m., your phone buzzes on the nightstand. Before you have said a single word, before you have made a single conscious decision, a cascade of invisible transactions has already begun. Your sleep-tracking app has logged eight hours and twelve minutes of rest, catalogued your movement patterns, measured your heart rate through the night, and transmitted this data your most intimate biological rhythms to servers you have never seen, owned by a company whose terms of service you clicked through in under three seconds three years ago.

You reach for the phone. You open Instagram. In the forty-five seconds you scroll before getting out of bed, Instagram's algorithm has noted the exact posts you lingered on, calculated the duration of each micro-pause, logged the content categories that slowed your thumb, and updated your psychographic profile. You are not browsing Instagram. Instagram is browsing you.

You get up. You ask Alexa for the weather. Amazon logs the timestamp, cross-references it with your purchase history, your browsing behavior, your Prime Video watch patterns, and adds one more data point to a profile that now spans years and contains millions of individual signals about who you are, what you want, and most valuably what you might be persuaded to do.

You check Gmail over breakfast. Google scans the content of every email. Not a human reading it, but an algorithm more thorough than any human could be extracting keywords, sentiment, relationships, financial signals, health indicators. Your inbox is a gold mine and you handed over the keys for free.

You drive to work. Google Maps tracks your route, your speed, your stops. It knows where you work, where you worship, where your children go to school, which medical offices you visit, which bars you frequent on Friday nights. This is not a side effect of the navigation service. This is the service. The turn-by-turn directions are the bait. Your location data is the catch.

By 9:00 a.m. before you have done a single minute of paid work you have been harvested dozens of times by systems you cannot see, by companies you barely think about, generating revenue from data you did not knowingly sell, under agreements you did not truly understand, feeding an industry worth trillions of dollars that is built entirely on one foundational premise:

You are not the customer. You are the product.

Understanding the Business Model That Runs Your Life

To understand why this matters why it is not merely a privacy annoyance but an existential threat to human autonomy we must first understand the economics driving it.

The surveillance economy is not an accident. It was not the unintended consequence of well-meaning technologists who simply got carried away. It was a deliberate architectural choice, made early in the life of the internet, that has compounded into a system of staggering scale and power. To see clearly how we got here, we need to go back to a moment of decision.

In the late 1990s and early 2000s, the internet faced a fundamental question: how do you sustain businesses built on giving things away for free? Email, search engines, social networks all of these provided enormous value to users who expected to pay nothing. The advertising model was the answer the industry settled on, but the advertising model of the internet era was fundamentally different from the advertising model of television or newspapers.

Traditional advertising was a blunt instrument. You bought airtime during a show watched by a particular demographic. You paid to reach an audience that mostly resembled the people you wanted to reach. The message was the same for everyone. The advertiser had no idea who specifically was watching, only a rough statistical picture of the likely audience.

Internet advertising changed everything. Because digital systems could track individual behavior which links you clicked, which pages you visited, how long you stayed, what you searched for advertisers could suddenly target specific people rather than statistical categories. And the more behavioral data a platform had about its users, the more precisely it could target, and the more it could charge advertisers.

This created an incentive structure with profound consequences: the more data you collect about people, the more money you make. Not somewhat more money. Dramatically more money. The difference between a generic ad shown to a random user and a hyper-targeted ad shown to someone whose exact desires, fears, and vulnerabilities have been mapped in detail is the difference between pennies and dollars.

The race was on. And the race has never stopped.

> **DATA POINT:** In 2023, Meta (Facebook's parent company) generated $131.9 billion in revenue. Approximately 98% of that revenue came from advertising. Meta's product is not a social network. Meta's product is the behavioral data of over 3 billion human beings, sold to anyone willing to pay for access to their attention.

The Architecture of Extraction

The machinery of corporate data harvesting is invisible by design. If you could see it if every data collection event generated a sound, a flash of light, a physical sensation you would not be able to function. The noise would be overwhelming. The lights would never stop flashing. The scale is simply incomprehensible to human intuition.

To make it comprehensible, let us break down the layers of the extraction architecture that surrounds every person with a smartphone and an internet connection in the modern world.

Layer One: The Apps You Choose

The most obvious layer is the apps you deliberately install and use. Facebook, Instagram, TikTok, Gmail, Google Maps, Spotify, Amazon these are the visible faces of the surveillance economy. You chose them. You opened accounts. You accepted terms of service. In this sense, there is an argument however thin that you consented to what they do with your data.

But consent as it exists in the digital economy is a fiction, and we will examine that fiction closely later in this chapter. For now, let us simply catalog what these apps collect:

Social media platforms collect every post, every comment, every like, every share, every message, every photo, every video, every search, every profile view, every advertisement you see and whether you interacted with it, the exact duration of every interaction, your location at the time of every action, the device you used, the network you were on, and dozens of other signals that you would never think to consider sharing.

Navigation apps collect your precise location continuously not just when you are actively navigating, but often in the background, building a continuous record of your physical movements through the world. Where you sleep. Where you spend your days. Where you go that you do not tell anyone about.

Retail apps and loyalty programs collect every purchase, building a detailed picture of your consumption habits, your income level, your health status (inferred from what you buy), your family composition, your political and religious affiliations (inferred from purchases), and the rhythms of your daily needs.

Health and fitness apps collect your most intimate biological data heart rate, sleep patterns, menstrual cycles, weight, exercise habits, stress levels. This data has particular value because it is both deeply personal and highly predictive of behavior and need.

Layer Two: The Trackers You Never Invited

The second layer of data extraction is invisible and uninvited. Embedded in nearly every website and app you use is a network of third-party trackers small pieces of code that do not belong to the site you are visiting but that piggyback on your visit to collect data for their own purposes.

These trackers operated by advertising networks, data brokers, analytics companies, and social media platforms follow you across the web. When you visit a news website, a recipe blog, a clothing retailer, and a travel booking site in the same afternoon, you may think you are having four separate, unrelated browsing sessions. But the tracking networks see all four visits as part of a single continuous portrait of you. They are connecting the dots across every site you visit, building a comprehensive picture of your interests, your intentions, and your vulnerabilities.

The tracker ecosystem has grown to monstrous proportions. Studies have found that the average website contains dozens of third-party trackers. Some contain hundreds. These trackers are operated by companies most users have never heard of companies like LiveRamp, Lotame, Oracle Data Cloud, Nielsen, and hundreds of smaller data brokers whose entire business is the collection, aggregation, and sale of behavioral data about individuals.

You did not consent to any of this. You did not even know it was happening.

FACT: A 2019 study by Ghostery found that the average website loaded 12.8 third-party trackers. Popular news sites regularly load 50 or more. Each of these trackers is collecting data about your visit and reporting it back to entities you have never interacted with.

Layer Three: The Data Brokers in the Shadows

Beneath the apps and beneath the trackers lies an industry that most people have never heard of but that knows more about them than their closest friends: the data broker industry.

Data brokers are companies that exist solely to collect, aggregate, enrich, and sell personal information. They sit at the center of a vast web of data flows, ingesting information from apps, trackers, public records, retail loyalty programs, financial institutions, health systems, government databases, and dozens of other sources, and combining it all into comprehensive profiles of individual people.

The profiles that data brokers maintain on individual Americans and on people around the world are staggering in their depth and detail. A typical data broker profile might include your name, address, and contact information; your family members and their information; your income and estimated net worth; your credit history; your medical conditions (inferred from purchases, searches, and app usage); your political affiliation; your religious beliefs; your sexual orientation; your relationship status; your employment history; your educational background; your criminal record if any; your social media activity; your online shopping history; your physical location patterns; and hundreds of other data points that together paint a portrait of your life more complete than anything you have ever written down about yourself.

This information is bought and sold constantly, in bulk, without your knowledge or meaningful consent, by an industry that operates in near-total obscurity. The largest data brokers companies like Acxiom, LexisNexis, Experian, Equifax, and TransUnion maintain files on hundreds of millions of people. Acxiom alone claims to have data on over 2.5 billion people worldwide.

Who buys this data? Advertisers, certainly. But also insurance companies assessing your risk profile. Employers conducting background checks. Political campaigns targeting voters. Law enforcement agencies building cases. Hedge funds analyzing consumer trends. Banks evaluating loan applications. Landlords screening tenants. And anyone else willing to pay which, given the low cost and high availability of this data, means almost anyone.

The Consent Illusion

At this point, a reasonable person might say: I clicked agree. I accepted the terms of service. If I did not want my data collected, I should have read the contract and made a different choice.

This argument has a surface plausibility that dissolves completely under examination. The notion that users meaningfully consent to digital data collection is one of the most pervasive and consequential myths in modern life. Let us dismantle it carefully.

The Terms of Service Problem

Researchers at Carnegie Mellon University calculated that if the average American were to read the privacy policies of every website and app they use, it would take approximately 76 work days per year. Not 76 hours. 76 full eight-hour work days. Every single year.

These policies are not written to be understood. They are written by teams of lawyers to provide legal protection to the companies that deploy them, with no meaningful expectation that any ordinary user will read them. They are dense, technical, and deliberately vague in the areas that matter most. They reserve the right to share your data with unspecified 'partners' and 'affiliates.' They allow data use for unspecified 'business purposes.' They can be changed at any time with minimal notice.

The typical privacy policy is a masterwork of deliberate obfuscation. Every clause that seems to protect you contains a loophole that swallows the protection whole. Every limit has exceptions broad enough to drive a truck through. The entire document exists not to inform you but to immunize the company against legal liability while appearing to do the former.

The Power Asymmetry Problem

Even if you read and understood every privacy policy which is not humanly possible you would still face a power asymmetry that makes meaningful consent impossible. The services collecting your data are not optional luxuries. They are, increasingly, the infrastructure of modern life.

You cannot participate in modern employment without email. You cannot navigate most cities without mapping apps. You cannot maintain social relationships with many people without social media. You cannot shop efficiently without online retail. You cannot access entertainment, news, or information without digital platforms. You cannot bank, communicate, or function in the modern economy without generating a constant stream of digital data.

Saying 'if you don't like it, don't use it' is like saying 'if you don't like the terms, don't use electricity.' The coercive power is not in the formal contract. It is in the impossibility of the alternative.

The Dark Patterns Problem

Beyond the unreadable contracts and the impossible alternative, the tech industry has perfected a set of manipulative design techniques called 'dark patterns' specifically engineered to steer users toward privacy-invasive choices.

When an app asks for your permission to track your location, it presents two options: 'Allow' in large, bright, easy-to-click text, and 'Don't Allow' in small, grey, hard-to-find text. When a platform offers privacy settings, the privacy-protective options are buried under multiple menus, worded in confusing technical language, and pre-set to the most invasive option by default.

The app that asks 'Would you like to allow Location Services?' relies on decades of research into human psychology the default effect, choice architecture, cognitive load to ensure that the vast majority of users will click yes without truly understanding what they are agreeing to. This is not user experience design. It is consent engineering, and it is practiced universally throughout the tech industry.

> **DOCUMENTED:** A 2022 Norwegian Consumer Council report found that Facebook's privacy settings were designed to nudge users toward sharing more data, using dark patterns including confusing language, strategic placement of options, and emotional framing. The report called the design 'privacy hostile by design.'

What They Know About You And What They Do With It

Abstract discussions of data collection can feel disconnected from real harm. Let us get specific. What does the surveillance economy actually know about individual people? And what, precisely, is done with that knowledge?

The Psychological Profile

Cambridge Analytica did not become infamous for nothing. What the Cambridge Analytica scandal revealed and what was deeply uncomfortable for the tech industry was how much behavioral data can reveal about the deepest recesses of a person's psychology.

Researchers at Cambridge University, working with Facebook 'like' data, demonstrated that they could predict a user's personality, political beliefs, sexual orientation, intelligence level, and susceptibility to various types of persuasion with accuracy that outperformed the users' own friends and family members. With 10 Facebook likes, they could predict personality more accurately than a coworker. With 150 likes, more accurately than a family member. With 300 likes, more accurately than a spouse.

Every major platform is doing this kind of psychological profiling, continuously and at massive scale. Your feeds, your recommendations, your advertisements are not random. They are the output of algorithms that have built a detailed model of your psychology and are using that model to show you the specific content and messaging most likely to produce the emotional reaction and behavioral outcome the platform desires.

This is not speculation. This is documented corporate behavior. Facebook's own internal research, revealed in the 2021 whistleblower documents leaked by Frances Haugen, showed that the company knew its algorithm was promoting emotionally inflammatory content particularly content that triggered outrage and anxiety because that content drove more engagement, which drove more time on platform, which drove more advertising revenue.

The algorithm did not care that the content was harmful. It was optimized for engagement, and outrage is engaging. Your psychological vulnerabilities were a resource to be exploited.

The Health and Insurance Implications

Your health data which you share constantly through fitness apps, period trackers, pharmacy apps, health search queries, and dozens of other channels is among the most sensitive and valuable data in the surveillance economy. It is also among the most dangerous when misused.

Insurance companies, pharmaceutical companies, and employers all have strong financial incentives to know about your health conditions, your health risks, and your health behaviors. Data brokers sell health-inferred data not your medical records, but data derived from your behavior that strongly suggests health conditions to anyone willing to pay.

A 2023 investigation by The Markup found that period-tracking apps were sharing intimate health data, including information about pregnancy, fertility, and sexual activity, with advertisers and data brokers. A 2022 study found that hospital websites were transmitting patient information including conditions being searched, appointment scheduling, and symptom checkers to Facebook through the Meta Pixel tracker, without patient knowledge or consent.

In a world where your health data is constantly leaking to commercial entities, the implications for insurance coverage, employment, and financial opportunity are profound and frightening. The protections of HIPAA apply to your doctor. They do not apply to the app on your phone.

The Financial Manipulation

Every financial decision you make online every price you see, every offer you receive, every loan rate you are quoted is potentially influenced by the data that companies have collected about you. This is not hypothetical. It is a documented and widespread practice called differential pricing or price personalization.

E-commerce platforms routinely show different prices to different users based on their inferred income, their purchase history, their location, their device, and their browsing behavior. A user browsing on a MacBook with a history of premium purchases may see higher prices than a user browsing on an entry-level Android. A user who has shown urgency checking the same product multiple times may see prices rise as the algorithm detects their intent to buy.

This is not just about paying more for goods online. Credit card companies use behavioral data to set credit limits and interest rates. Insurance companies use data to set premiums. Employers use data to evaluate candidates. Landlords use data to evaluate tenants. At every financial interaction, the asymmetry of information created by surveillance capitalism works against the individual and in favor of the institutional actor holding the data.

The Political Targeting

Data-driven political targeting is not new. But the scale and sophistication of what is now possible and what is being practiced represents a qualitative change in the relationship between political power and individual citizens.

Political campaigns, political action committees, and critically foreign state actors use commercial behavioral data to identify psychologically vulnerable voters, map their susceptibility to specific emotional appeals, and deliver precisely tailored messaging designed not to inform or persuade through rational argument but to trigger emotional responses that drive behavior at the ballot box.

The tools used are the same tools used to sell you sneakers. The data is the same data that shows you advertisements. The difference is the stakes: instead of purchasing decisions, the targeted behavior is voting, civic engagement, and political identity.

And the foreign adversaries who have demonstrated interest in manipulating Western democracies Russia, China, Iran, and others have access to the same commercial data markets that domestic actors use. The behavioral data that American citizens have been generating for decades is available for purchase by anyone with a credit card and a data broker account. The surveillance economy does not discriminate between domestic advertisers and hostile foreign intelligence operations.

The Companies Doing It A Field Guide

It is easy to speak abstractly about 'tech companies' and 'the surveillance economy.' It is more useful and more alarming to look specifically at the major actors and examine what they actually do.

Meta: The Social Graph as Surveillance Infrastructure

Meta the parent company of Facebook, Instagram, and WhatsApp is the most sophisticated behavioral surveillance operation in human history. With over 3 billion monthly active users across its platforms, Meta has constructed what may be the most detailed map of human social relationships, behaviors, and psychology ever assembled.

Meta tracks users not just on its own platforms but across the broader web through the Meta Pixel a piece of tracking code embedded in over 10 million websites that reports your browsing behavior back to Meta whether you are logged into Facebook or not. Meta has admitted to using this data to build what it calls 'shadow profiles' of people who have never created a Facebook account.

Meta's business depends entirely on the depth and accuracy of its user profiles. The more it knows about you your relationships, your interests, your emotional states, your purchasing behavior, your political and religious beliefs the more it can charge advertisers to reach you. The incentive to collect more, know more, and use more never stops pushing in one direction.

Facebook's own researchers have repeatedly found that the platform's algorithm promotes content that makes users unhappy, angry, and anxious because unhappy, angry, anxious users spend more time on the platform seeking resolution or validation. The company has known this for years and has consistently chosen engagement metrics over user wellbeing.

Google: The Search Engine as Mind Reader

Google processes over 8.5 billion search queries per day. Each search query is a window into the human mind a direct expression of need, curiosity, fear, desire, and intention that people would never share voluntarily with another person but type freely into a search box because it feels private.

It is not private. Google stores every search query, associates it with your account or device, and uses it to build a profile of your interests, concerns, and psychology that updates in real time. Google's advertising system which generates the vast majority of its revenue allows advertisers to target users based on their search history, their YouTube viewing history, their Gmail content, their Google Maps location history, their Google Shopping behavior, and their activity on websites that use Google Analytics and Google Ads tracking.

The scale of Google's data operation is impossible to fully comprehend. Google sees a substantial portion of all activity on the internet, through its own platforms (Search, YouTube, Gmail, Maps, Chrome, Android) and through its third-party tracking infrastructure (Google Analytics, Google Tag Manager, DoubleClick). When you use the internet, you are almost certainly generating data for Google regardless of whether you are on a Google product.

Android the operating system running on roughly 72% of the world's smartphones is a data collection platform. The phone in your pocket is, among other things, a Google data collection device that you carry with you everywhere and that reports your location, your app usage, your contacts, your communications, and your behavior continuously to Google's servers.

Amazon: The Store That Knows Your Life

Amazon began as a bookstore. It is now the most comprehensive commercial surveillance operation in the consumer economy, with a reach into people's homes, behaviors, and consumption patterns that no retailer in history has approached.

Amazon knows what you buy. It knows what you searched for and did not buy. It knows what price would have convinced you to buy. Through the Alexa smart speaker now present in tens of millions of homes Amazon has a listening device in people's most private spaces, recording ambient audio and building voice profiles. Through Ring doorbell cameras now deployed on an estimated one in ten American homes Amazon has constructed a private neighborhood surveillance network, with footage shared with law enforcement on request.

Amazon Web Services, the company's cloud computing division, powers a substantial fraction of the internet's infrastructure. Companies that compete with Amazon's retail operation store their customer data on Amazon's servers. The company's visibility into the commercial internet is extraordinary and largely unexamined.

Amazon's advertising business now one of the largest digital advertising platforms in the world uses purchase data to target advertising with a precision that other platforms envy. If you searched for a product on Amazon and did not buy it, you will see advertisements for that product or its competitors across the web for days. Your commercial intentions are an asset that Amazon monetizes constantly.

The Data Broker Industry: The Invisible Layer

Behind the consumer-facing tech giants lies an entire industry that most people have never heard of and that operates with almost no public accountability. Data brokers companies like Acxiom, LexisNexis Data Solutions, Oracle Data Cloud, Experian Marketing Services, and hundreds of smaller operators are the wholesale market of the surveillance economy.

These companies do not interact with consumers. They interact with data. They buy it from apps, retailers, financial institutions, public records agencies, and other sources. They combine it, clean it, enrich it, and sell it to anyone willing to pay. They are the connective tissue of the surveillance economy, enabling any actor with sufficient funds to access a comprehensive profile of virtually any American adult.

The data broker industry operates with minimal legal oversight in most of the world. In the United States, there is no comprehensive federal law governing data brokers. They are not required to register with any government authority. They are not required to disclose what data they hold on individuals. They are not required to allow individuals to see, correct, or delete their data though some states have begun to require this.

The result is a shadow economy of personal information that operates entirely outside the awareness of the people it concerns.

The Legal Vacuum How Corporations Got Away With It

The question of how the surveillance economy was allowed to develop to its current scale without meaningful legal constraint is not a simple one. It involves a combination of legislative failure, regulatory capture, ideological assumptions, and the sheer speed at which technology outpaced governance.

The American Approach: Self-Regulation That Never Regulated

Unlike Europe, which developed a comprehensive legal framework for data protection culminating in the General Data Protection Regulation of 2018 the United States took a sectoral and largely voluntary approach to data privacy. Different industries have different rules: healthcare data is protected by HIPAA, financial data by the Gramm-Leach-Bliley Act, children's data by COPPA. But there is no overarching federal privacy law that gives Americans comprehensive rights over their personal data.

For decades, the dominant regulatory philosophy was that the market would solve the problem. If consumers valued their privacy, they would choose products that protected it. Companies that abused data would lose customers. Self-regulation would emerge because businesses had reputational incentives to behave responsibly.

This philosophy was completely wrong. It ignored the information asymmetry that made informed consumer choice impossible, the addictive design of digital platforms that made switching costs prohibitively high, and the network effects that meant the most privacy-invasive platforms were also the ones with the most social value. The market did not protect privacy. The market devoured it.

The Lobbying Industrial Complex

Every time serious federal privacy legislation has been introduced in Congress, the tech industry has deployed its considerable lobbying resources to prevent, dilute, or indefinitely delay it. The tech sector spends hundreds of millions of dollars annually on lobbying, making it one of the most powerful political forces in Washington.

The lobbying arguments follow predictable patterns: privacy regulation will stifle innovation, hurt small businesses, harm the free internet, cost jobs, and reduce consumer choice. These arguments are made by the same companies that simultaneously argue that consumers have voluntarily chosen to share their data and have nothing to complain about. The contradiction if consumers freely choose data sharing, why does mandating transparency threaten business models? is never addressed.

The result is a regulatory environment that has consistently failed to keep pace with technological reality. While Europe enacted GDPR and California passed its own comprehensive privacy law, the federal government has spent decades holding hearings, expressing concern, and doing nothing.

Section 230 and the Shield of Immunity

Section 230 of the Communications Decency Act 26 words buried in a 1996 telecommunications law has done more to shape the internet than perhaps any other piece of legislation. It provides platforms with immunity from liability for content posted by their users, which was a reasonable policy in 1996 when the internet was young and platforms had limited capacity to monitor user content.

But Section 230 has been interpreted so broadly that it protects platforms not just from liability for user content but from liability for their own algorithmic choices the decisions about what content to amplify, what content to suppress, and what to show to whom. The algorithm that promotes rage-inducing content because it drives engagement is protected. The design choice to bury privacy settings under multiple menus is protected. The decision to target emotionally vulnerable users with manipulative advertising is largely protected.

The combined effect of inadequate privacy law and broad Section 230 immunity has created a legal environment in which the surveillance economy can operate with near-total impunity. The companies that built this system did not do so in spite of the law. They did so in the space the law created for them.

The Psychological Cost What Surveillance Does to Human Beings

The harms of the surveillance economy are not merely financial or political. There is a profound psychological cost to living in a world where your every action is watched, recorded, analyzed, and monetized. Understanding this cost is essential to understanding why data privacy is not a niche technical concern but a fundamental human issue.

The Chilling Effect

Surveillance changes behavior. This is not a theory it is one of the most robust findings in social psychology. When people know they are being watched, they conform more closely to perceived social norms, take fewer risks, and suppress behaviors, beliefs, and expressions they fear may be judged negatively. This is known as the chilling effect.

In the context of digital surveillance, the chilling effect is pervasive and consequential. Research has found that Americans changed their internet search behavior after the Snowden revelations about NSA surveillance avoiding searches on sensitive topics related to terrorism, health, sexuality, and political dissent. If people know their searches are watched, they self-censor their intellectual curiosity.

People share less, explore less, and express themselves less freely when they know they are under surveillance. In a free society, the right to private thought to entertain ideas, explore questions, and form opinions without observation is foundational to intellectual freedom. The surveillance economy systematically erodes that right.

The Manipulation of Attention and Emotion

The business model of most major digital platforms depends on maximizing user attention. Time on platform equals advertising revenue. Every design choice every notification, every infinite scroll, every algorithmically curated feed is engineered to capture and hold attention as long as possible.

The tools of attention capture are not neutral. They exploit known psychological vulnerabilities: the variable reward mechanism that makes social media feeds as compelling as slot machines, the social validation loop that makes likes and comments as neurologically rewarding as cocaine, the fear of missing out that drives compulsive checking behavior. These are not accidents of design. They are intentional features, developed through extensive A/B testing on hundreds of millions of users, to maximize the addictive quality of the platform.

The psychological literature on heavy social media use is consistent and damning. It is associated with increased rates of depression, anxiety, loneliness, body image problems, sleep disruption, and decreased life satisfaction particularly among adolescents. The companies that built these platforms know this. Their own internal research documents it. And they have consistently chosen engagement metrics over user mental health.

The Loss of Authentic Self

When people know that their behavior is being recorded and analyzed when they know that their social media posts, their purchases, their search history, and their physical movements are all being catalogued and used to build a profile there is a natural tendency toward self-performance. You become, to some degree, the character that your data suggests, curating your digital footprint to project the image you want the algorithms to see.

This has subtle but significant effects on identity and authenticity. When your digital self is a product that is constantly being valued and sold, the boundary between who you are and who your data says you are begins to blur. Young people especially who have grown up with social media as a constant presence may struggle to develop authentic selves apart from the performance of selfhood that digital platforms demand.

This is not a small thing. The development of authentic identity, the freedom to explore and fail and evolve without a permanent record, the right to be inconsistent and contradictory and human these are not luxuries. They are the conditions for a fully realized human life. The surveillance economy systematically undermines them.

The Scale of the Problem Numbers That Should Frighten You

Abstract arguments about surveillance are important. But sometimes the sheer scale of the problem, rendered in numbers, communicates what words alone cannot.

> **$** The global digital advertising market the economic engine of the surveillance economy was valued at approximately $740 billion in 2023. By 2028, it is projected to exceed $1 trillion.
>
> **USERS:** Google has approximately 4.3 billion users. Meta has over 3 billion. Amazon serves hundreds of millions of customers. Apple has over 2 billion active devices. The combined surveillance reach of these four companies alone encompasses virtually the entire adult population of the developed world.
>
> **DATA POINTS:** Acxiom, one of the largest data brokers, claims to hold an average of 1,500 individual data points on each of the more than 700 million people in its database.
>
> **BREACHES:** Over 353 million individuals were affected by data breaches in the United States in 2023 alone a number that has roughly tripled since 2020. Every breach is a leak from the surveillance economy, exposing data that never should have been collected to actors who should never have access to it.
>
> **DARK PATTERNS:** A 2022 study by the Norwegian Consumer Council found dark pattern designs in 97% of

the top apps analyzed, all designed to steer users toward the most data-invasive choices.

These numbers are not background statistics. They are measurements of a system that touches every person reading this page. The surveillance economy is not something that happens to other people, in other countries, in some abstract digital realm. It is happening to you, right now, constantly, at a scale that is difficult to fully grasp.

The Path Forward Reclaiming Your Data Identity

This chapter has been, by design, alarming. The surveillance economy is genuinely alarming. But alarm without agency is just anxiety. The question that matters is not merely how bad this is it is what we can do about it.

We will explore specific, actionable steps in detail in the final chapter of this book. But it is worth closing Chapter One with a foundational reorientation a shift in how we think about our relationship to the digital platforms that have come to dominate our lives.

The Mindset Shift: From User to Sovereign

The first and most important step is conceptual. We need to stop thinking of ourselves as users and start thinking of ourselves as sovereigns. Your data is not a byproduct of your digital activity. It is a representation of your life your relationships, your thoughts, your health, your movement through the world. It belongs to you in a way that should be as inviolable as any other form of personal property.

When you understand your data as an expression of your personhood rather than a technical artifact of platform use, the question changes. It is no longer 'how do I navigate the privacy settings of this platform?' It is 'why should any entity have access to this expression of my personhood without my explicit, informed, and freely given consent?'

This is a radical question in the context of the current digital economy. The entire economic model of the surveillance industry depends on people not asking it. When enough people ask it when enough people demand that their data be treated as property they own rather than a resource others extract the political and economic calculus changes.

Immediate Actions You Can Take Today

While we await broader systemic change, there are concrete steps every individual can take to reduce their exposure to the surveillance economy:

Use a privacy-focused browser. Firefox with privacy extensions or the Brave browser blocks many of the third-party trackers that follow you across the web. The default settings of Chrome are optimized for Google's data collection, not your privacy.

Use a private search engine. DuckDuckGo, Brave Search, and Startpage do not build behavioral profiles of their users. Switching your default search engine is one of the highest-leverage privacy changes most people can make.

Audit your app permissions. Go through the apps on your phone and review what permissions they have been granted location, microphone, camera, contacts, calendar. Revoke any permissions that are not genuinely necessary for the app's core function.

Use a VPN. A Virtual Private Network hides your internet traffic from your Internet Service Provider itself a significant data collector and from many web-based trackers. Not all VPNs are equally trustworthy; choose one with a verified no-logging policy.

Opt out where you can. California residents have legal rights to opt out of data sales. Many data brokers offer opt-out procedures. These are tedious and must be periodically repeated, but they meaningfully reduce your exposure. Services like DeleteMe automate much of this process.

These are not complete solutions. The systemic problem requires systemic solutions regulation, legislation, and corporate accountability that go beyond what any individual can achieve through personal behavior. But they are starting points. And starting points matter.

The Bigger Stakes

We began this chapter with a mundane morning coffee, a phone check, a drive to work. We end it here, with a question that is anything but mundane:

What kind of society do we want to live in?

A society in which every person's thoughts, feelings, relationships, health, and physical movements are continuously monitored, recorded, and sold is not a free society in any meaningful sense. It is a society in which freedom exists as long as no one with access to your data decides to use it against you. It is a society in which autonomy is contingent on the goodwill of corporations and whoever buys their data.

That is not the society most people believe they live in. And it is not the society that the architects of modern democracy imagined when they wrote about liberty, privacy, and the rights of individuals against institutional power. The surveillance economy has built, in plain sight, an infrastructure of control that would have horrified them.

Reclaiming privacy is not about having something to hide. It is about insisting on the conditions necessary for a free life the freedom to think, to explore, to err, to grow, to be human without every moment of that humanity being extracted and monetized by entities that do not have your interests at heart.

That insistence is not paranoia. It is not technophobia. It is, at its core, the oldest and most fundamental claim of free people against those who would control them:

My life belongs to me.

* * *

END OF CHAPTER ONE

Continue to Chapter Two: The Government You Didn't Vote For

CHAPTER TWO

The Government You Didn't Vote For:

State Surveillance, the Typhoon Hackers, and the War Already Being Fought Inside America's Walls

---

"The PRC cyber threat is not theoretical. What we've found to date is likely the tip of the iceberg." CISA Director Jen Easterly, Congressional Testimony, January 2024

The Intruders Who Were Already Inside

Imagine waking up one morning to find that someone has been living in your house for two years.

They have not stolen anything. They have not broken anything. They have moved nothing. They have left no trace visible to the naked eye. But they have been there, quietly, in the walls and beneath the floors, watching. They have memorized every room. They know which windows lock and which ones don't. They have photographed your daily schedule, catalogued the contents of every drawer, mapped every circuit breaker in the electrical panel. They have not acted. Not yet. They are waiting for the right moment the moment when you are most vulnerable, when disruption will cause the most damage, when there is the least you can do to stop it.

Now imagine that your house is not just your home. It is the water system that serves your city. It is the power grid that keeps hospitals running. It is the communications infrastructure that connects your military. It is the financial network that processes every transaction in the country.

This is not a metaphor. This is what happened to the United States of America.

In the years between 2021 and 2024, agents of the Chinese government operating under the codename Volt Typhoon silently infiltrated American critical infrastructure on a scale that U.S. intelligence officials have described as unprecedented in the history of cyberwarfare. They did not come in loud. They did not announce themselves with ransomware screens or dramatic data thefts. They came in quietly, using the legitimate tools of the networks they invaded, leaving almost no trace, and they stayed for months, for years burrowing deeper, mapping everything, waiting.

At the same time, a second operation Salt Typhoon was unfolding inside America's telecommunications backbone. The phone calls of sitting U.S. senators. The text messages of a presidential campaign. The wiretapping systems used by federal law enforcement to conduct court-authorized surveillance. All of it, compromised. All of it, read by operators reporting to China's Ministry of State Security.

And the American public was not told.

Not for months. Not until the scale of the intrusions had reached such proportions that silence was no longer possible. Not until October 2024, when the story finally broke not from a government press conference, not from a transparent briefing to the American people, but from investigative reporters who had pieced together the truth from fragments and leaks.

This chapter is about what that truth means. Not just as a cybersecurity story though it is certainly that but as a story about power, sovereignty, and the government that operates in the dark spaces between what you are told and what is actually happening.

You did not vote for the Chinese Communist Party. But for years, it has had access to systems that affect your daily life, your security, and the functioning of the nation you live in. And the government that was supposed to protect you from that access was, in many cases, too slow, too fragmented, and too politically constrained to stop it.

That is the government you didn't vote for. And it has been governing your life in ways you never knew.

Understanding the Typhoon Family: China's Cyber Army

Before we can grasp the scale of what has happened, we need to understand the actors involved. The Chinese government's cyber offensive apparatus is not a single organization with a single mission. It is a coordinated ecosystem of state-sponsored hacking groups, each with distinct targets, techniques, and objectives, all operating under the strategic direction of the People's Republic of China's military and intelligence services.

The naming convention that has emerged from Western cybersecurity firms and government agencies is the 'Typhoon' family a series of threat actor designations assigned primarily by Microsoft and adopted by U.S. government agencies. Each 'Typhoon' represents a distinct Chinese state-sponsored operation. Three of them Volt Typhoon, Salt Typhoon, and Flax Typhoon have been at the center of what U.S. intelligence has described as the most alarming sustained cyberattack campaign in American history.

Understanding each of them, what they targeted, how they operated, and what they were trying to achieve, is essential to understanding both the threat the United States faces and the profound failures that allowed that threat to take root.

Volt Typhoon: The Phantom in the Infrastructure

Volt Typhoon is, in the assessment of U.S. federal agencies, the most strategically dangerous of the Chinese hacking operations currently operating against American targets. Its mission is not espionage in the traditional sense. It is not primarily interested in stealing secrets or harvesting communications. Its purpose is preparation for war.

FBI Director Christopher Wray, testifying before the U.S. House Select Committee on the Strategic Competition Between the United States and the Chinese Communist Party on January 31, 2024, called Volt Typhoon 'the defining threat of our generation.' That is an extraordinary statement not diplomatic hedging, not bureaucratic caution, but a direct, unambiguous declaration that a foreign state actor had positioned itself to cause catastrophic harm to the United States. It deserves to be taken seriously as exactly what it is.

What makes Volt Typhoon uniquely frightening is not its sophistication alone though its technical capabilities are formidable but its patience and its stated intent. CISA, NSA, and the FBI, in a joint advisory released in February 2024, assessed with high confidence that Volt Typhoon actors were pre-positioning themselves on U.S. critical infrastructure IT networks specifically to enable the disruption or destruction of critical services in the event of a major crisis or military conflict with the United States.

Read that sentence again. These hackers are not in American systems to steal. They are there to wait. They are positioning themselves now so that when not if, in their planning, but when a conflict between China and the United States erupts over Taiwan or any other flashpoint, they can throw a switch and plunge America into chaos.

> **CONFIRMED:** U.S. authoring agencies including CISA, NSA, and FBI confirmed that Volt Typhoon compromised multiple critical infrastructure organizations primarily in the Communications, Energy, Transportation Systems, and Water and Wastewater Systems sectors in both the continental United States and its territories including Guam.

The technique Volt Typhoon uses to achieve this near-invisible presence is called 'living off the land' (LOTL). Rather than installing malware that might be detected by security software, Volt Typhoon actors use the legitimate, built-in tools of the networks they infiltrate. They log in with valid credentials. They run native system commands. They move through networks using the same tools that administrators use every day. To automated security systems and to human analysts reviewing logs, their activity looks like normal network traffic because they have engineered it to be indistinguishable from normal network traffic.

CISA has documented instances of Volt Typhoon actors maintaining persistent access inside victim networks for at least five years without detection. Five years. Half a decade of presence inside American infrastructure inside the systems that pump water to cities, route electricity to hospitals, and move trains through metropolitan areas without triggering a single alert significant enough to prompt removal.

In one chilling documented case, Volt Typhoon maintained unauthorized access to a Massachusetts power utility's operational technology network for nearly a year from February to November of 2024 mapping the electrical grid's structure, learning its operational procedures, and collecting data that would enable future disruption. The utility, Littleton Electric Light and Water Departments, eventually detected and contained the breach, but only after months of undetected presence.

The utility in Massachusetts was small and obscure. If Volt Typhoon was patient enough to spend almost a year inside a minor municipal power utility, what has it been doing and for how long inside the major systems that serve tens of millions of Americans?

Salt Typhoon: The Phone Call You Thought Was Private

While Volt Typhoon was burrowing into the physical infrastructure of America, a second operation was unfolding at the level of personal communications and its targets included some of the most powerful people in the United States.

Salt Typhoon, which has been active since at least 2019 and is attributed to China's Ministry of State Security, executed what cybersecurity experts have called one of the most severe telecommunications hacks in history. Beginning in at least 2023 and continuing through its public disclosure in late 2024, Salt Typhoon infiltrated the networks of nine of America's largest telecommunications companies including AT&T, Verizon, T-Mobile, Lumen, Spectrum, Consolidated Communications, and Windstream companies whose combined infrastructure carries the private communications of virtually every American adult.

The scope of what Salt Typhoon accessed was staggering and took months to fully understand a process that, according to U.S. officials, is still ongoing. Among the documented intrusions:

Salt Typhoon accessed metadata records showing who called whom, when, for how long, and from where of more than one million Americans, most of them located in the Washington D.C. area. In the intelligence community, metadata is often more valuable than content. Knowing who a person communicates with, when, and how frequently, reveals networks of relationships, identifies assets and sources, and maps the structure of organizations.

In some cases, Salt Typhoon obtained not just metadata but the actual content of phone calls and text messages. The targets of direct content access included staff members of the Kamala Harris 2024 presidential campaign and phones belonging to then-President-elect Donald Trump and Vice President-elect J.D. Vance. The phones of the incoming and outgoing leaders of the United States were being listened to by Chinese intelligence.

Most alarmingly, Salt Typhoon penetrated the private portals that telecommunications companies are legally required to maintain for law enforcement access the backdoors through which the FBI, NSA, and other agencies conduct court-authorized wiretapping of investigation targets. By accessing these systems, Salt Typhoon potentially obtained a list of every person or organization under active U.S. government surveillance, including Chinese intelligence assets and informants who were being monitored by American counterintelligence. If those sources were burned if China now knows who we were watching the damage to American intelligence operations is incalculable.

CONSEQUENCE: By penetrating U.S. law enforcement surveillance portals, Salt Typhoon may have given China the ability to identify American intelligence sources and assets,

> effectively turning America's own surveillance infrastructure into a tool for Chinese counterintelligence.

In December 2024, at a private meeting in Geneva, Chinese officials were reported to have all but admitted to being behind the Volt Typhoon campaign linking it explicitly to U.S. support for Taiwan. It was as close to a confession as the diplomatic world ever produces. The message was unmistakable: this is what we will do if you stand in our way over Taiwan. The critical infrastructure intrusions are not reconnaissance. They are a loaded gun, pointed at American society, with a finger on the trigger.

Flax Typhoon and the Army National Guard

A third operation Flax Typhoon targeted a different layer of American defense. Linked to PRC information security companies that operate under direct government direction, Flax Typhoon focused on Taiwan's government and technology sectors as well as U.S. critical infrastructure both domestically and abroad. Flax Typhoon created a botnet a network of compromised devices used to mask the origin of attacks from hundreds of infected Internet of Things devices, providing operational cover for ongoing operations.

But perhaps the most alarming disclosure associated with the Typhoon campaign family came from a Department of Homeland Security memo in 2024, which revealed that Salt Typhoon had extensively compromised a U.S. state's Army National Guard network over nine months in 2024 from March through December. The breach, which DHS declined to specify by state, resulted in the exfiltration of administrator credentials, detailed network traffic diagrams, a geographic map of National Guard installations throughout the state, and the personally identifiable information of National Guard service members.

Administrator credentials. Network diagrams. The physical locations of military installations. The identities of citizen soldiers. In a single breach of a single state's National Guard network, China obtained an operational blueprint for disrupting military response capability at the state level precisely the kind of capability that would be critical in the opening hours of any conflict that required military mobilization inside the United States.

And DHS's assessment was direct: if Salt Typhoon successfully exploited this data, it could hamstring state-level cybersecurity partners' ability to defend U.S. critical infrastructure against Chinese cyber campaigns in any future crisis.

The Five-Year Blind Spot: How We Failed to See

The most terrifying aspect of the Typhoon revelations is not what China did. It is how long China was able to do it before anyone noticed.

Five years of persistent access. Years of undetected presence inside the most critical systems in the United States. A sustained, coordinated, sophisticated campaign against American infrastructure that evaded detection by the federal agencies, private companies, and intelligence services whose entire mandate was to prevent exactly this.

How does that happen? The answer is not simple, and it is deeply uncomfortable, because it implicates not just technical failures but structural, political, and philosophical failures that go to the heart of how America approaches cybersecurity.

The Fragmentation Problem

The United States has no single, unified cyber defense architecture. Responsibility for cybersecurity is scattered across dozens of federal agencies CISA, NSA, FBI, CYBERCOM, DHS, DOE, and many others each with its own mandate, its own authorities, its own relationships with the private sector, and its own jealously guarded turf. Coordinating a response to a sustained, sophisticated, multi-vector attack across this fragmented landscape is extraordinarily difficult under the best circumstances.

Critical infrastructure in the United States is primarily owned and operated by private companies. The power grid, the telecommunications networks, the water systems, the financial networks these are not government assets. They are corporate assets, operated for profit, with cybersecurity budgets determined by boards of directors who weigh security spending against quarterly earnings.

The federal government can offer guidance, can issue advisories, can conduct voluntary assessments but it cannot compel private companies to maintain the security standards necessary to defend against a sophisticated state-sponsored adversary.

Volt Typhoon exploited this gap relentlessly. It targeted the edges of networks the routers, VPN appliances, and firewalls operated by small vendors and municipal utilities that sit at the boundary between the corporate internet and the operational technology systems that control physical processes. These edge devices are often old, often unpatched, often forgotten. They are the cybersecurity equivalent of leaving a window unlocked in a house you thought was fully alarmed.

The Living Off the Land Problem

Traditional cybersecurity defense is built around the assumption that attackers will use tools and techniques that can be identified as malicious malware that can be detected by antivirus software, traffic patterns that can be flagged as anomalous by intrusion detection systems, commands that do not match normal user behavior.

Volt Typhoon's living-off-the-land technique destroyed this assumption entirely. By using only legitimate, native tools Windows Management Instrumentation, PowerShell, Remote Desktop Protocol, network diagnostic commands Volt Typhoon generated activity that was, by definition, indistinguishable from normal administrative behavior. The only way to detect it was through meticulous behavioral analysis over long periods, comparing each action against a detailed baseline of what normal looked like for that specific network.

Very few organizations even well-resourced federal agencies maintain that level of behavioral monitoring. The monitoring infrastructure required is expensive, the analysis is labor-intensive, and the expertise required is scarce. Volt Typhoon was not just technically sophisticated. It was strategically sophisticated it targeted the specific gap between what most organizations' security programs were designed to detect and what was actually happening.

The Classification Problem

When the U.S. government began to piece together the scope of Volt Typhoon and Salt Typhoon, much of what it learned was classified. Intelligence about Chinese operations, their methods, and their access was protected at the highest levels meaning it could not be shared with the private sector companies whose networks were compromised without elaborate security clearance processes that most corporate IT staff do not hold.

This created a surreal situation: the federal government knew that specific companies had been compromised, knew how the compromise had been achieved, but could not tell those companies in enough detail to help them find and remove the intruders, because the information was classified. The bureaucracy of secrecy, designed to protect national security, was itself a barrier to national security.

CISA Director Jen Easterly acknowledged this problem directly in congressional testimony, noting that the agency was working to find ways to share threat information with private sector partners without compromising classified sources. But the structural tension between classification and defense has not been resolved, and it represents a fundamental vulnerability in America's cyber defense architecture that sophisticated adversaries have learned to exploit.

> **CISA DIRECTOR JEN EASTERLY, 2024:** The PRC cyber threat is not theoretical. CISA teams have found and eradicated Volt Typhoon intrusions into critical infrastructure across multiple sectors. And what we've found to date is likely the tip of the iceberg.

What They Were Preparing For: Taiwan and the Cyber First Strike

To understand the strategic logic of Volt Typhoon's infrastructure pre-positioning, you need to understand the scenario that Chinese military planners have been gaming for decades: a conflict over Taiwan.

Taiwan the Republic of China is a self-governing democracy of 23 million people that the People's Republic of China claims as its own territory and has never renounced the use of force to 'reunify.' The United States, under a policy of 'strategic ambiguity' formalized in the Taiwan Relations Act of 1979, has committed to providing Taiwan with the means to defend itself and has left deliberately unclear whether American military forces would directly intervene in a Chinese military action against the island.

For Chinese military planners, a conflict with the United States over Taiwan is not a distant hypothetical. It is the central scenario around which China's military modernization and its cyber warfare strategy has been organized for years. The question is not whether such a conflict might happen, but what China can do to maximize its chances of success if and when it does.

The answer, in the cyber domain, is to pre-position devastating capabilities inside American critical infrastructure before any shots are fired. The logic is coldly rational: if the United States decides to intervene militarily in defense of Taiwan, it will require the mobilization of naval forces, logistical supply chains, communications networks, and the political will of an American public that must be prepared to accept the costs of conflict. A cyber first strike executed at the moment conflict begins aimed at disrupting the power grid, contaminating the water supply, paralyzing transportation systems, and cutting communications could slow or prevent American military mobilization while simultaneously shattering the public will to fight.

The Guam Scenario

Guam holds a special place in China's strategic calculus and in Volt Typhoon's targeting pattern. The small U.S. territory in the Western Pacific is home to Andersen Air Force Base and Naval Base Guam two of the most critical American military installations in the Indo-Pacific region, from which any American military response to a Taiwan conflict would almost certainly be launched.

Volt Typhoon's documented targeting of Guam's telecommunications and critical infrastructure was not accidental. Microsoft, in its May 2023 advisory that first publicly identified Volt Typhoon, noted explicitly that the group could 'disrupt critical communications infrastructure between the United States and Asia region during future crises.' FBI Assistant Director Brett Leatherman confirmed in early 2025 that Volt Typhoon was specifically focused on critical infrastructure centered around the U.S. Navy, particularly in island communities like Guam.

Disabling Guam's communications infrastructure at the outset of a Taiwan conflict would not just inconvenience American forces. It would blind them severing the command-and-control links that allow the U.S. military to coordinate a response across the vast distances of the Pacific theater. It would be the cyber equivalent of cutting the telephone lines before the invasion begins. And Chinese cyber actors have been quietly working toward the capability to do exactly that for years.

The Infrastructure Attack Scenario

Beyond Guam, Volt Typhoon's documented presence in American water systems, power grids, and transportation networks suggests a broader playbook for a cyber opening to any conflict. Consider what coordinated disruption of these systems would mean for the American home front in the event of a conflict with China:

Water system disruption. Volt Typhoon has documented access to water and wastewater treatment systems across the United States. Compromising water treatment controls could force the shutdown of water supplies to major population centers or, in the most alarming scenarios, allow manipulation of treatment processes in ways that could affect public health. Cities without clean water are not just inconvenienced they are incapacitated, their populations in crisis, their governments overwhelmed.

Power grid disruption. Electric utilities confirmed compromised by Volt Typhoon include systems serving millions of Americans. The modern United States cannot function without electricity. Hospitals, communications networks, financial systems, fuel distribution, emergency services all depend on a functioning grid. A sustained, targeted power outage affecting major population centers or key military logistics hubs would create cascading crises that would overwhelm emergency response capabilities and divert enormous attention and resources from any military mobilization.

Transportation disruption. Rail systems, port operations, and air traffic control systems have all been targeted by Chinese cyber actors. The logistics of American military mobilization depend critically on civilian transportation infrastructure. Disrupting rail freight, shutting down port operations, or creating chaos in air traffic control does not just harm civilians it directly impedes the military's ability to move personnel and equipment.

Communications disruption. Salt Typhoon's deep penetration of telecommunications infrastructure creates the capability to selectively disrupt or monitor communications not just for intelligence purposes, but for active disruption when conflict begins.

The combination of these capabilities all positioned, all waiting, all under the control of operators who have been inside these systems for years represents something genuinely new in the history of warfare. The battlefield has been laid out not in some distant theater but inside the borders of the United States itself, in the infrastructure that Americans depend on every day.

> **STRATEGIC ASSESSMENT:** Chinese officials in a December 2024 Geneva meeting reportedly acknowledged the Volt Typhoon campaign was related to U.S. support for Taiwan. The pre-positioning of destructive cyber capabilities inside American critical infrastructure is not espionage. It is preparation for coercion.

Salt Typhoon's Human Cost: Real People, Real Consequences

It is tempting to treat cyber intrusions as abstract a battle of algorithms and protocols fought in the invisible realm of digital networks, remote from the lived experience of ordinary people. But Salt Typhoon's telecommunications hack was not abstract. It had real human consequences for real people, and those consequences illuminate the stakes of the surveillance war in a way that technical descriptions cannot.

The Presidential Candidates' Phones

In October 2024, the FBI and CISA confirmed that Salt Typhoon had compromised the telecommunications infrastructure used by both major presidential campaigns. The phones of Donald Trump and J.D. Vance the Republican presidential and vice presidential nominees had been accessed by Chinese intelligence. Staff members of the Kamala Harris campaign had also been targeted.

Let that sink in for a moment. In the final weeks of an American presidential election, the leading candidates for the nation's highest office were communicating over lines that a foreign government was listening to. The strategy sessions, the debate preparations, the private conversations between campaign officials and donors and advisors all potentially accessible to operators working for China's Ministry of State Security.

The FBI issued urgent security advisories to both campaigns. CISA recommended that senior government officials and political figures switch immediately to end-to-end encrypted applications Signal, FaceTime, iMessage and abandon regular phone calls and text messages for any sensitive communication. The agency that exists to protect

American communications infrastructure was telling Americans, in essence, do not trust American communications infrastructure.

The broader implication was chilling: if the presidential candidates' communications were compromised, whose were not? Every senator, every general, every intelligence official, every diplomat who was using standard telephone and SMS communications was potentially exposed. The working assumption had to be that China was listening.

The Wiretapping System Breach

Of all the revelations in the Salt Typhoon story, none is more strategically significant or more alarming in its implications than the confirmed breach of the Communications Assistance for Law Enforcement Act portals.

CALEA, passed in 1994, requires telecommunications companies to build and maintain technical infrastructure that allows law enforcement to conduct court-authorized surveillance of specific phone numbers and accounts. These are the systems through which the FBI serves a wiretap order on a carrier and the carrier enables real-time access to the target's communications. They are also the systems through which U.S. intelligence agencies monitor foreign targets operating inside the United States.

Salt Typhoon accessed these systems. By penetrating the CALEA portals of multiple major carriers, Chinese intelligence potentially obtained visibility into which individuals and organizations the United States government was actively surveilling. This is a counterintelligence catastrophe of the first order.

Consider the implications. If China knows who the FBI is wiretapping which of its intelligence officers, which of its informants and assets operating in the United States, which of its front companies it can alert those targets, modify their behavior, feed misinformation into the surveillance channels, or simply eliminate the problem. Years of patient, careful American counterintelligence work could be neutralized overnight.

Intelligence officers refer to this kind of damage as 'burning' a program or an asset. When a source is burned when a foreign intelligence service learns that their operative is being watched the source is compromised, the intelligence stream is cut off, and the counterintelligence value of years of work evaporates. Salt Typhoon may have burned an unknown number of American intelligence programs simultaneously, across multiple carriers, in a single sustained operation.

The full scope of what was accessed which targets, which investigations, which intelligence programs has not been publicly disclosed and may never be. But officials who have been briefed on the damage assessments have used words like 'catastrophic' and 'historic' that leave little room for optimism.

The Treasury Department Breach

While the telecommunications hacks were unfolding, a separate Salt Typhoon operation was targeting the financial levers of American government power. In early 2024, Salt Typhoon breached the U.S. Department of the Treasury's Departmental Offices network through a compromised third-party cloud service provider BeyondTrust that was supporting Treasury's network operations.

The targets inside Treasury were not random. Salt Typhoon specifically accessed the systems of the Office of Foreign Assets Control the division responsible for administering and enforcing U.S. economic sanctions. OFAC is one of the most powerful tools in the American foreign policy arsenal: it can cut companies, individuals, and entire countries off from the U.S. financial system, with devastating economic consequences.

By accessing OFAC's systems, China potentially obtained advance knowledge of pending sanctions actions a capability that would allow sanctioned entities to move assets, restructure financial relationships, and otherwise evade the consequences of American economic pressure before the sanctions took effect. The element of surprise that makes sanctions effective would be gone. The Treasury Department's most powerful financial weapon would be blunted before it was even drawn.

The devices of former Treasury Secretary Janet Yellen, Deputy Secretary Wally Adeyemo, and Acting Undersecretary Brad Smith were reportedly among those compromised. The people who set American economic policy who decide which individuals and entities get cut off from the global financial system were being watched by the government against which that economic policy is most heavily directed.

> **SANCTIONED:** The Treasury Department sanctioned Sichuan Juxinhe Network Technology Co. and Shanghai-based hacker Yin Kecheng in January 2025 for their direct roles in Salt Typhoon operations. The company was identified as having strong ties to China's Ministry of State Security.

Domestic Surveillance: The Government That Watches Its Own

The Chinese cyber threat is the most dramatic and concrete example of state-level surveillance threatening American citizens. But it is not the only one, and a complete account of the government you did not vote for must also reckon with the surveillance that operates closer to home.

The revelations of Edward Snowden in 2013 exposed, to a stunned global public, the scope of the National Security Agency's domestic data collection programs that swept up the phone records of millions of Americans, the communications of allied foreign leaders, and the data of major internet companies on a scale that had never been publicly acknowledged. The legal and political fallout from those revelations was significant but ultimately limited. The programs were partially reformed, partially curtailed, partially renamed and the fundamental architecture of domestic digital surveillance largely persisted.

This chapter is not primarily about NSA surveillance of Americans that is a complex topic that has been extensively documented elsewhere. But the context matters: the infrastructure of surveillance that Salt Typhoon compromised the CALEA wiretapping portals, the telecommunications metadata systems exists because American law enforcement and intelligence agencies built it for their own use. The backdoors that Chinese intelligence walked through were doors that the U.S. government required telecommunications companies to build.

This is the deep, uncomfortable irony at the heart of the surveillance debate: the legal mandates that require communications companies to maintain surveillance access for law enforcement create precisely the infrastructure vulnerabilities that foreign adversaries exploit. You cannot build a backdoor that only the good guys can use.

A door is a door, and if it exists, someone will eventually find a way through it.

The FISA Court and the Secret Law

The Foreign Intelligence Surveillance Act established a secret court the FISA Court to authorize surveillance of foreign intelligence targets operating in or connected to the United States. The court operates entirely in secret. Its rulings which have, over time, developed a body of legal precedent about the scope of permissible surveillance were classified. Americans were subject to a body of law they could not read, interpreted by a court they could not observe, authorizing surveillance they did not know about.

The concept of secret law is antithetical to the foundational principles of democratic governance. The rule of law requires that the law be knowable that citizens can understand the rules by which they are governed and hold their government accountable for following them. The FISA Court represents a substantial departure from that principle, justified by national security imperatives that the public is not permitted to evaluate.

The Section 702 program which authorizes collection of communications involving non-Americans that pass through U.S.-based systems and which has been repeatedly reauthorized with minimal reform has been documented to collect substantial amounts of American citizens' communications as incidental collection. The exact scope of this 'incidental' collection has never been publicly disclosed.

The Data Broker Pipeline to Government

As we explored in Chapter One, the commercial data broker industry has built comprehensive profiles on virtually every American adult. What we did not fully explore there is the relationship between that commercial surveillance infrastructure and U.S. government surveillance.

Federal agencies including law enforcement agencies and intelligence services routinely purchase commercial data from data brokers to conduct investigations without obtaining the court orders that the Fourth Amendment would require if they collected the same data directly. The logic is straightforward: if a private company has the data and will sell it, and if there is no law requiring a warrant to purchase commercial data, the government can bypass Fourth Amendment protections by simply buying what it otherwise could not legally seize.

The FBI, DHS, ICE, and other agencies have all been documented using commercial data broker services to obtain location data, communications metadata, social network analysis, and other information about individuals under investigation without warrants. The Supreme Court's 2018 decision in Carpenter v. United States suggested some limits on this practice by requiring warrants for extended cell-site location history, but the broader practice of government data broker purchases exists in a largely unresolved legal gray zone.

The same data that tech companies harvest and sell to advertisers your location history, your social connections, your behavioral patterns, your financial situation is also available to any government agency with a budget line for commercial data purchases. You did not consent to becoming a surveillance subject of your own government. But the infrastructure that was built to sell you advertisements has made you one.

China's Social Credit System: A Warning from the Future

To understand where the convergence of government surveillance and digital data infrastructure can lead in the absence of meaningful democratic constraint, look at what China has built.

China's Social Credit System a term that encompasses multiple distinct programs at the national and local level uses behavioral data, financial data, legal records, biometric surveillance, and AI-driven analysis to assign individuals and companies a score that determines their access to services, travel, employment, and social opportunity. Individuals with low scores can be barred from purchasing airline or train tickets, blocked from attending certain schools, denied loans, publicly named as 'untrustworthy,' and subjected to enhanced monitoring.

The system does not require a conviction. It does not require due process. It requires only that an algorithm, fed by the surveillance infrastructure, decide that your behavior falls below an acceptable threshold. The standard is not legality it is conformity. The system punishes not just those who break the law but those who dissent, those who criticize, those who associate with the 'wrong' people.

Western commentators have often treated the Social Credit System as a uniquely Chinese phenomenon, alien to democratic societies. This comfort is misplaced. The underlying components behavioral surveillance, data aggregation, algorithmic scoring, automated consequence are already present in Western commercial and governmental systems. Credit scores, insurance algorithms, predictive policing models, hiring AI, and benefits eligibility systems already use behavioral data to make consequential decisions about people's lives without meaningful transparency or accountability.

The difference between China's Social Credit System and the systems currently operating in Western democracies is not a difference in kind. It is a difference in degree and in the explicit authorization of government use. That gap is narrowing.

The Political Response: Too Little, Too Late, Too Divided

How did the United States government respond to the Typhoon revelations? The answer, in aggregate, is: inadequately, inconsistently, and in ways that reflect the profound dysfunction of the American political system when confronted with threats that require sustained, bipartisan commitment.

The Delayed Disclosure

The timeline of public disclosure is itself a damning document. Volt Typhoon's activities in Guam were first publicly identified by Microsoft in May 2023. The FBI disrupted part of Volt Typhoon's botnet infrastructure in January 2024. A joint CISA/NSA/FBI advisory warning of the threat to critical infrastructure was issued in February 2024. Yet the American public whose water systems and power grids were the targets was never told in plain, accessible terms what had happened and what it meant.

Salt Typhoon's telecommunications breaches were discovered in the autumn of 2024 and publicly disclosed in October 2024 but only because investigative reporters broke the story. The government's initial acknowledgment was grudging and partial, revealing the minimum necessary to manage the public revelation rather than proactively informing citizens of the threat to their communications.

By December 2024, the Cyber Safety Review Board the independent body established after major cyber incidents to investigate and report on the government's response had begun a formal investigation of the Salt Typhoon breaches. Then the incoming Trump administration fired all members of the board before it could complete its investigation. The report that might have provided the most comprehensive public accounting of what happened, how it happened, and what the government knew and when, was never completed.

The Sanctions That Came Late

The Treasury Department's sanctioning of Sichuan Juxinhe Network Technology and hacker Yin Kecheng in January 2025 in the final days of the Biden administration was a meaningful step but one that came years after the operations it was responding to had already achieved their objectives. Sanctioning the operators after the breach provides some accountability and some deterrent signal, but it does not undo the access that was obtained, the intelligence that was collected, or the pre-positioning that was accomplished.

Deterrence in the cyber domain is fundamentally different from deterrence in conventional military domains. You cannot uninvade an infrastructure. You cannot un-exfiltrate stolen data. By the time sanctions are levied, the damage is already done. The logic of consequence after the fact, applied to operations that take years to plan and execute, offers limited protection.

The Political Weaponization of Cybersecurity

Cybersecurity policy in the United States has been increasingly infected by partisan politics in ways that directly impair the country's ability to defend itself. The dismissal of the Cyber Safety Review Board by the incoming Trump administration before it could complete its Salt Typhoon investigation was widely interpreted as political, given that the investigation would have produced a report covering breaches that occurred during the Biden administration.

But the politicization of cybersecurity is not a one-party phenomenon. Every administration has had areas where political considerations constrained cybersecurity responses where imposing costs on adversaries risked diplomatic complications that were deemed more important, where acknowledging vulnerabilities would require admitting failures, where funding security required spending money that other constituencies wanted for other purposes.

The result is a cybersecurity posture that lurches between administrations, lacks strategic continuity, and consistently fails to maintain the sustained, long-term investment that defending against sophisticated state actors requires. China's cyber operations have been running continuously for years with consistent strategic direction. America's response has been episodic, politicized, and institutionally fragmented.

> **WARNING:** As of February 2026, Senator Maria Cantwell had written to Committee Chairman Ted Cruz requesting a hearing in which the CEOs of AT&T and Verizon would be compelled to account for Salt Typhoon. Nine months after carriers claimed to have contained the breach, fundamental questions about network security remain publicly unresolved.

The Average American in the Crossfire

At this point, a reader might reasonably ask: this is all very alarming at a national level, but what does it mean for me specifically? I am not a senator or a presidential candidate. My phone calls are not of interest to Chinese intelligence. Why should the Typhoon revelations change how I think about my own security?

The answer unfolds on several levels, from the immediately personal to the broadly societal.

Your Communications Are Infrastructure

Salt Typhoon did not target only the communications of the politically powerful. It accessed the metadata of over one million Americans, the majority of them in the Washington D.C. area but not limited to government officials. The telecommunications infrastructure it compromised carries every American's communications.

When CISA issued its emergency guidance recommending that Americans switch to encrypted messaging apps, it was not recommending this only for senators and generals. It was recommending it for anyone who values the privacy of their communications which should be everyone. The guidance was addressed to 'highly targeted individuals' but acknowledged that the same security practices are appropriate for all users.

If the underlying telecommunications infrastructure is compromised, every call, every text, every metadata record generated on that infrastructure is potentially accessible to the attacker. You may not be an intelligence target. But your communications pass through the same systems that intelligence targets use, and the access that Salt Typhoon obtained was systematic, not surgical.

Your Daily Life Depends on the Infrastructure Under Attack

Volt Typhoon's targets are not abstract national security assets. They are the systems that keep you alive and functional on a daily basis. The water that comes out of your tap. The electricity that powers your home, your workplace, your hospital. The trains and traffic systems you navigate. The communications networks that connect you to emergency services.

A coordinated disruption of these systems of the kind that Volt Typhoon has been pre-positioning to execute would not be a political or military event that unfolds on television while you go about your business. It would be a direct assault on your daily existence. Days without clean water. Nights without power. Medical emergencies unable to reach hospitals that cannot operate. Financial systems unable to process transactions. Supply chains unable to deliver food.

These are not hypothetical worst cases invented for dramatic effect. They are the scenarios that American intelligence agencies, speaking in classified briefings to congressional overseers, have assessed as plausible consequences of the access that Volt Typhoon has already obtained. The pre-positioning has been accomplished. The capability exists. Whether it is used depends on decisions made by the Chinese government in response to geopolitical events that none of us fully control.

Your Data Is a Target

The personal data that the commercial surveillance economy has accumulated about you your location history, your medical information, your financial behavior, your social connections, your political and religious beliefs is not just a commercial asset. It is an intelligence asset, and foreign governments treat it as such.

China has been engaged in systematic, large-scale theft of American personal data for years. The 2015 Office of Personnel

Management breach not a Typhoon operation but a related Chinese intelligence operation resulted in the theft of security clearance applications for over 21 million current and former federal employees. These applications contain extraordinarily detailed personal information: financial history, foreign contacts, psychological evaluations, past drug use, sexual history, family members. The information was a treasure trove for Chinese intelligence, enabling the identification and potential blackmail or recruitment of American government employees.

The Anthem health insurance breach in 2015, attributed to Chinese state-sponsored actors, exposed the personal health and financial information of approximately 79 million Americans. The Marriott hotel breach in 2018, also attributed to Chinese actors, exposed the passport numbers, contact information, and travel histories of approximately 500 million hotel guests.

Your personal data the data you generate daily through your phone, your apps, your purchases, your healthcare is being harvested not just by corporations seeking to sell you things, but by foreign intelligence services seeking to understand, influence, and in some cases compromise you. The surveillance economy that Chapter One described as a threat to your autonomy is also, simultaneously, a national security vulnerability of the first order.

The Broader Pattern: China Is Not Alone

It would be a mistake a comfortable but dangerous mistake to frame the state-level cyber threat to American citizens as exclusively a Chinese problem. China is the most sophisticated, most persistent, and most strategically alarming adversary in the cyber domain. But it is not the only one, and understanding the broader landscape of state-sponsored cyber operations against American interests is essential to understanding the full scope of the threat.

Russia: The Disruptive Force

Russia's cyber operations against the United States have a different character from China's. Where China tends toward patient, long-term strategic operations focused on pre-positioning and intelligence collection, Russia has demonstrated a greater appetite for disruptive operations designed to create visible chaos and undermine American democratic institutions.

The SolarWinds hack of 2020 in which Russian intelligence compromised a widely used IT management software platform and used it to gain access to the networks of thousands of government agencies and private companies was a masterclass in supply chain compromise that anticipated many of the techniques later used in Volt Typhoon's living-off-the-land operations. The breach affected the Treasury Department, the State Department, the Department of Homeland Security, and dozens of other agencies. It was, in the words of then-Senator Mark Warner, 'the most sophisticated cyberattack in history.'

Russia's interference operations in the 2016 and subsequent American elections demonstrated a different but related capability: the use of cyber operations not to disrupt infrastructure but to disrupt democracy itself hacking campaign systems, stealing and selectively releasing information, and using social media manipulation to amplify division.

The combination of Russia's willingness to act disruptively and China's patient, strategic pre-positioning creates a threat environment in which American critical infrastructure and democratic institutions face sustained pressure from multiple sophisticated state actors simultaneously.

Iran: The Escalating Adversary

Iran's cyber capabilities, while less sophisticated than China's or Russia's, have grown substantially over the past decade, and Iran has demonstrated a willingness to conduct attacks against American targets that cross lines that other adversaries have generally respected.

The 2021 attempt to increase sodium hydroxide concentrations in a Florida water treatment plant attributed to Iranian state-sponsored actors was a rare instance of a state-sponsored cyber operation targeting an attack that could have directly harmed civilian lives. The operation was detected before it succeeded, but it demonstrated that at least one state actor was willing to use cyber access to water infrastructure for purposes that went beyond pre-positioning.

Iran has also targeted American financial institutions, critical infrastructure operators, and government agencies, and has conducted aggressive information operations targeting American political processes. The convergence of Iranian operational willingness with increasingly sophisticated capabilities represents a growing threat that is sometimes obscured by the focus on Chinese and Russian operations.

North Korea: The Criminal State

North Korea's cyber program represents a unique threat profile: a state actor that uses cyber operations primarily for financial gain, stealing cryptocurrency and bypassing international sanctions through digital theft at a scale that funds a significant portion of its weapons programs.

North Korea's Lazarus Group has been linked to billions of dollars in cryptocurrency theft, targeting exchanges, DeFi platforms, and individual investors.

The convergence of state intelligence capabilities and criminal financial motivation in a single operation is particularly difficult for law enforcement and intelligence agencies to counter, as it blurs the distinction between criminal jurisdiction and national security response.

The Surveillance State You Consented To: TikTok and the Data Sovereignty Question

Any chapter on government surveillance and the digital sovereignty of American citizens must confront the most contentious and revealing data-sovereignty debate of recent years: TikTok.

TikTok, the short-form video platform operated by the Chinese company ByteDance, has over 170 million users in the United States. It is one of the most powerful social media platforms ever created its algorithm is, by the consensus of the technology industry, the most effective content recommendation system in consumer history, capable of identifying and delivering content that maximizes engagement with a precision that its competitors have struggled to match.

The national security concern with TikTok is twofold. First: ByteDance is a Chinese company, subject to the laws of the People's Republic of China, including the National Intelligence Law of 2017, which requires Chinese companies and individuals to support and cooperate with state intelligence work. Under this law, any data that ByteDance holds about American users could theoretically be compelled to be turned over to Chinese intelligence.

Second: TikTok's algorithm which decides what 170 million Americans see when they open the app could theoretically be used as an influence operation tool, promoting or suppressing content in ways that serve Chinese government interests. The distinction between organic content moderation and state-directed information manipulation is not verifiable from outside the company.

The U.S. Congress passed legislation requiring ByteDance to divest TikTok's American operations or face a ban. The Supreme Court unanimously upheld the law in January 2025. The subsequent months have seen political maneuvering that has delayed implementation, and the company's ultimate disposition remains unresolved at the time of this writing.

The TikTok debate is a microcosm of the broader data sovereignty challenge. The question is not only whether a specific company can be trusted with American data. It is whether the legal and regulatory framework that governs the collection, storage, and use of American citizens' data provides meaningful protection against foreign government access and the honest answer is that it does not.

The data that American companies collect about American citizens under the minimal constraints of current American privacy law is traded, sold, and accessible in ways that create ample opportunity for foreign government access without the dramatic front-door intrusions represented by the Typhoon operations. The back door is the entire commercial data ecosystem. Every data broker file, every behavioral profile, every location history sold to advertisers is also potentially available, through purchase or theft, to any foreign intelligence service with the resources and the interest to obtain it.

The Future Threat: What the Next Decade Holds

If the revelations of 2024 were alarming and they should have been alarming the trajectory of the threat is more alarming still. The capabilities that Volt Typhoon and Salt Typhoon have demonstrated represent the current state of the art in state-sponsored cyber operations against critical infrastructure and communications systems. What is coming in the next decade is worse.

AI-Accelerated Cyber Operations

Artificial intelligence is transforming the speed, scale, and sophistication of cyber operations in ways that will fundamentally change the threat landscape within years. AI systems can automate the reconnaissance and exploitation phases of cyber attacks, identifying vulnerabilities and crafting attack approaches faster than human teams ever could. They can generate convincing phishing communications customized to individual targets, adapt attack techniques in real time in response to defensive measures, and analyze the vast quantities of stolen data communications records, credential databases, behavioral profiles with a thoroughness and speed that would require armies of human analysts.

The living-off-the-land techniques that made Volt Typhoon so difficult to detect depended on patient human operators carefully mimicking normal behavior to avoid triggering security alerts. AI-driven operations will be able to do this continuously, at scale, across thousands of compromised systems simultaneously, adapting in real time to defensive responses with a speed and consistency that human operators cannot match.

The defenders will also have AI tools. But the structural advantages in this competition favor the attacker: an attacker needs to find one way in, a defender must protect every possible entry point. AI amplifies both sides of this equation, but its amplification of offensive capability is, in the assessment of most cybersecurity experts, at least as great as its amplification of defensive capability.

Quantum Computing and the Encryption Crisis

Every communication that is currently encrypted every secure website, every encrypted email, every VPN tunnel, every secure military communication is encrypted using mathematical algorithms that are computationally infeasible to break with current classical computing hardware. The security of these systems rests on the fact that it would take longer than the age of the universe to brute-force the encryption keys.

Quantum computing changes this. A sufficiently powerful quantum computer running an algorithm known as Shor's Algorithm can factor the large prime numbers that underlie current encryption standards in a matter of seconds rather than geological epochs. When quantum computing reaches sufficient scale, the encryption that protects virtually all digital communications and financial systems will be broken.

Intelligence agencies, including the NSA, have assessed that a cryptographically relevant quantum computer one capable of breaking current encryption standards may be achievable within the next decade. China is investing massively in quantum computing research and has demonstrated significant advances in quantum communication systems.

The implication is that encrypted communications that are being intercepted and stored today the vast surveillance databases that intelligence agencies of all countries maintain will be retroactively decryptable when quantum capability arrives. Communications that are encrypted now may be read in a decade. This practice store now, decrypt later is a known intelligence strategy, and it means that the value of current encryption protections has a time limit.

Governments and standards bodies are working to develop post-quantum cryptography encryption algorithms resistant to quantum attack. The transition to these new standards will take years and will require the replacement or upgrade of essentially all digital security infrastructure. Whether that transition happens fast enough to stay ahead of quantum capability is one of the defining cybersecurity questions of the coming decade.

The 5G and 6G Battlefield

The global competition to build and control next-generation telecommunications infrastructure 5G and the emerging 6G standards is not merely a commercial contest. It is a geopolitical battle over which government will have the greatest visibility into the communications infrastructure on which the global digital economy depends.

Huawei and ZTE, the Chinese telecommunications equipment manufacturers, were effectively banned from U.S. networks on national security grounds precisely because of concerns that Chinese law would require the companies to provide the Chinese government with access to data passing through their equipment. The same concern applies, with varying degrees of urgency, to Huawei infrastructure deployed in allied and partner countries whose communications intersect with American national security interests.

As 5G deployment continues and 6G standards are being defined, the architecture decisions being made now will determine the surveillance and access capabilities built into global telecommunications infrastructure for the next generation. The government that shapes those standards, and whose companies build the equipment, will have structural advantages in the intelligence and cyber operations of the coming decades.

What Individuals Must Understand and What They Can Do

This chapter has covered ground that is genuinely frightening. State-sponsored cyber actors have penetrated American critical infrastructure. Presidential communications have been intercepted. The systems that safeguard law enforcement surveillance have been compromised. The data that defines your digital existence is an intelligence asset sought by foreign governments. Quantum computing threatens to retroactively expose encrypted communications.

The response to genuine fear is not paralysis. It is understanding, action, and advocacy. Let us address each.

Understanding Your Exposure

The first step is clarity about what the actual threat to ordinary individuals looks like, as distinct from the national security threats described above.

For the vast majority of Americans, the direct threat from Chinese state cyber operations is not that your phone calls will be individually targeted for interception. It is that the infrastructure that carries your communications has been compromised in ways that could affect everyone who uses it. It is that the data you have generated through your digital life is available to foreign intelligence services through commercial and illicit channels. It is that the critical infrastructure that keeps your daily life functioning has foreign actors pre-positioned inside it.

These are genuine threats, but they operate at the infrastructure level more than the individual level for most people. The exception is individuals in certain categories: government employees with security clearances, defense contractor personnel, researchers in sensitive technical fields, journalists, activists, political campaign staff, and business executives in sectors of strategic interest to foreign governments. If you fall into any of these categories, your individual exposure is significantly higher than average, and you should take personal operational security seriously as a professional obligation.

Practical Steps for Everyone

Switch to end-to-end encrypted communications. CISA recommended this explicitly in response to Salt Typhoon. Signal is the gold standard it uses the most rigorously audited encryption protocol available, collects minimal metadata, and has been audited repeatedly by independent security researchers. For phone calls, Signal's voice calling is preferable to regular phone calls. For messaging, Signal is preferable to SMS.

Understand that SMS text messages are not secure. This cannot be overstated in the post-Salt Typhoon environment. SMS messages travel in the clear through telecommunications carrier infrastructure. That infrastructure has been demonstrated to be accessible to foreign intelligence. Treat SMS as a postcardnot a sealed letter.

Use strong, unique passwords and a password manager. This sounds basic because it is basic. But an enormous proportion of successful credential-based intrusions including many of those used by Volt Typhoon to gain initial access to infrastructure networks exploit reused or weak passwords. A password manager generates and stores strong, unique credentials for every service, eliminating the most common class of credential vulnerability.

Enable multi-factor authentication everywhere. Two-factor authentication, particularly using an authenticator app rather than SMS-based codes, adds a critical layer of protection against credential theft. CISA has made this a primary recommendation in all of its public guidance on individual security.

Be aware of the data you share with apps, particularly those with connections to foreign governments. This does not mean living in paranoid digital isolation. It means understanding that an app's country of origin matters, that data collected by apps is subject to the laws of the country in which the company operates, and that the convenience of any particular application should be weighed against the sensitivity of the data it accesses.

Stay informed. The public deserves to know what is happening to the infrastructure their lives depend on. The Typhoon revelations were not disclosed proactively by a government committed to public transparency they were forced out by investigative journalism and congressional pressure. Pay attention to credible cybersecurity reporting. Understand that your government's cyber vulnerabilities are your vulnerabilities.

The Advocacy Imperative

Individual security measures are necessary but not sufficient. The structural vulnerabilities that allowed Volt Typhoon and Salt Typhoon to operate for years inside American systems cannot be fixed by individuals switching to Signal. They require institutional responses: sustained government investment in critical infrastructure security, mandatory security standards for essential systems, meaningful information sharing between government and the private sector, and political will to prioritize cybersecurity as the national security issue it demonstrably is.

That political will is generated by an informed and engaged citizenry. Every voter who understands the stakes of the Typhoon revelations and demands accountability from their elected representatives is contributing to the political conditions necessary for meaningful reform. Every journalist who covers these issues with rigor and depth is performing an essential public service. Every business leader who invests seriously in security rather than minimizing it as a cost center is strengthening the overall infrastructure.

The government that operates in the dark spaces the one that failed to disclose the Typhoon intrusions promptly, that dismantled its own investigative board before it could report, that has consistently underinvested in cyber defense relative to the scale of the threat is not an inevitable fixture. It is the product of political choices made by people who respond to political pressure. That pressure begins with citizens who know what is happening and refuse to accept the comfortable fiction that someone else is handling it.

The War Already Being Fought

We began this chapter with an image: someone living in your walls for two years, learning everything, waiting for the right moment.

The truth is that the image is already obsolete. It is not one intruder. It is a coordinated operation involving hundreds of skilled operators, backed by the resources of a nation-state with a multi-trillion dollar economy, executing a long-term strategic plan that has been in development for years and is aimed at achieving Chinese dominance in the Indo-Pacific region and, if necessary, at the cost of American power and American lives.

The war that China is preparing for is not being fought with ships and missiles not yet. It is being fought with packets and protocols, with credentials and compromised edge devices, with patient reconnaissance of electrical substations and water treatment plants and telecommunications backbones. It is being fought in the digital infrastructure of everyday American life, by an adversary that has committed to the long game, that operates with strategic continuity and national purpose, and that has already demonstrated capabilities that American defenses have been far too slow to counter.

This is the government you didn't vote for. Not just China's government which has been making decisions about American infrastructure without your knowledge or consent but your own government, which has known about these intrusions for years, disclosed them inadequately, responded to them inconsistently, and allowed the political process to compromise the continuity of cyber defense.

You deserve better from both. And the first step toward demanding it is understanding what is actually happening.

What is actually happening is a cyberwar. It is already underway. The battlefield is the infrastructure that keeps you alive. And you are standing on it.

* * *

END OF CHAPTER TWO

Continue to Chapter Three: The AI Impersonation Crisis

CHAPTER THREE

The AI Impersonation Crisis: *When Your Face, Voice, and Identity Are No Longer Yours*

"I mean, the picture of him it was him." Steve Beauchamp, 82-year-old retiree, after losing $690,000 of his life savings to a deepfake Elon Musk video

The Voice on the Phone

Sharon Brightwell was doing the dishes when her phone rang.

It was her daughter. She recognized the voice instantly the cadence, the slight Florida accent, the way she always started a sentence before she was quite ready. But this time her daughter was crying. Sobbing, really, the kind of crying that makes a mother's stomach drop before her brain has even processed the words. Through the tears came a story: a car accident, a miscarriage, a run-in with the law. She needed help. She needed it now. A lawyer was on the line. He needed fifteen thousand dollars in cash, today, before the situation became irreversible.

Sharon did not hesitate. That was her daughter's voice. She could hear it. She drove to the bank. She withdrew the cash. She handed it to a courier.

She never spoke to her daughter during any of this. Her daughter was fine. Her daughter was at work, unaware that her voice scraped from a few seconds of audio on her Instagram page, processed through an AI voice-cloning tool that costs nothing and takes minutes to learn had been used to rob her mother of fifteen thousand dollars in a single afternoon.

Sharon Brightwell is not a foolish woman. She is not technologically naive in any remarkable way. She is a mother who heard her child's voice in distress and responded the way mothers are supposed to respond. The people who built the tool that stole her daughter's voice counted on exactly that. They understood that the most powerful security system in the human mind the recognition of a loved one's voice could be defeated with fifteen seconds of audio and freely available software.

And they were right.

This is the AI impersonation crisis. It is not a future problem. It is not a Hollywood thought experiment. It is happening right now, to real people, at a scale and velocity that is accelerating faster than almost any threat in the history of consumer fraud. It is happening to grandmothers in Florida and finance workers in Hong Kong and voters in New Hampshire and employees of billion-dollar corporations everywhere. It is happening because a technology of extraordinary capability and almost zero barrier to access has been deployed into a world with almost no legal protection for the people it victimizes.

Almost no protection, that is, for ordinary people.

Because there is a different story unfolding in parallel a story about powerful people who looked at the same technology, recognized the same threat, and moved with remarkable speed and determination to protect themselves. Hollywood actors. Recording artists. Billionaires. Politicians. The wealthy, the famous, and the institutionally connected saw the AI impersonation crisis coming for their faces and their voices and their reputations, and they organized. They lobbied. They struck. They negotiated contracts. They wrote laws.

They got protected.

And the question that burns at the center of this chapter the question that should make every ordinary American furious is this: if the threat is serious enough that Tom Hanks needs a law to protect his face, why does Sharon Brightwell not deserve the same protection for her daughter's voice?

The answer is not that she does not deserve it. The answer is that she does not have the power to demand it. Not yet. But understanding how the powerful protected themselves understanding the mechanics of their outrage and their response is the first step toward demanding that the same protections be extended to everyone.

What Deepfakes Are and Why They Are Different

The word 'deepfake' has entered common usage, but its full technical and social implications are not always understood. Before we can grasp the scope of the crisis, we need to understand precisely what this technology is, how it works, and why it represents a qualitative departure from previous forms of deception and impersonation.

A deepfake is a piece of synthetic media video, audio, or both generated by artificial intelligence systems that have been trained to replicate a specific person's appearance, voice, and mannerisms. The term derives from the deep learning techniques specifically generative adversarial networks and, more recently, diffusion models and large language models that power the most sophisticated versions of the technology.

The key word is replicate. Not imitate, not approximate, not suggest. Replicate. Modern deepfake technology, when properly deployed, does not produce content that looks or sounds like a person. It produces content that is neurologically indistinguishable from that person. Your brain, processing the audio or video signal, receives inputs that are identical in their relevant characteristics to the inputs it would receive if the person were actually speaking. The signal that tells your brain 'this is real' fires just as it would for genuine content. The cognitive safeguards that have evolved over hundreds of thousands of years to identify authentic human communication are simply not calibrated to detect this threat. They were not designed to. Nothing in human evolutionary history required them to be.

The Technology Has Left the Building

For most of the history of deepfake technology roughly from 2017, when the term was coined in a viral Reddit post, through approximately 2022 the creation of convincing deepfakes required substantial computational resources, significant technical expertise, and hours of work. The technology was concerning but its high barrier to entry limited its deployment to well-resourced actors: state intelligence agencies, organized criminal groups, and occasionally technically sophisticated individuals with specific targets.

That era is over.

The democratization of AI tools has placed professional-grade deepfake capability in the hands of anyone with a laptop and an internet connection. Tools like ElevenLabs, Speechify, Resemble AI, and dozens of others allow voice cloning from as little as fifteen seconds of sample audio. Tools like DeepFaceLive, Magicam, and Amigo AI allow real-time face replacement during live video calls, meaning that you can appear on a video conference as virtually any person whose image is available online, in real time, without post-production.

The barrier to entry has not just dropped. It has effectively disappeared.

An investigative report by 404 Media documented in 2024 that scammers were using generative AI tools to alter their face, voice, gender, and race during live video calls in real time making video verification, which many organizations had adopted as a response to audio phishing, essentially useless as a security measure. A tool that was designed to confirm visual identity had been defeated by a technology that can replicate visual identity in real time on consumer hardware.

> **SPEED OF ACCESS:** It takes approximately 8 minutes and a few dollars in compute costs to create a convincing voice deepfake using commercially available tools. The voice samples required can be scraped from a public social media account. No technical expertise is required.

The Scale of the Assault

The numbers documenting the growth of deepfake fraud are among the most alarming statistics in contemporary crime reporting, and they are accelerating in ways that suggest not a trend but a transformation.

Between 2022 and 2023, deepfake fraud cases worldwide increased by 1,740% in North America alone, according to data compiled by identity verification firm Sumsub. In 2024, deepfake attacks were occurring at a rate of approximately one every five minutes globally. In the first half of 2025, deepfake-related fraud incidents reached 580 nearly four times the total number recorded in all of 2024. Global losses from deepfake-enabled fraud reached over $200 million in the first quarter of 2025 alone, and cumulative losses by mid-2025 had exceeded $897 million.

The Deloitte Center for Financial Services has projected that generative AI-enabled fraud in the United States will reach $40 billion annually by 2027 a number that represents a compound annual growth rate of 32% from the $12.3 billion recorded in 2023. These are not worst-case projections. They are central-case estimates based on current trajectories.

By 2027, AI-enabled fraud will cost the United States alone more than the entire annual GDP of many small nations. Every year. Recurring. Accelerating.

The Celebrity Awakening: When the Powerful Realized They Were Targets

The AI impersonation crisis became a mainstream political issue not when ordinary people started losing their life savings though they had been losing them for months before anyone in power paid serious attention but when famous, powerful, and wealthy people discovered that the technology threatened them personally.

This is not a comfortable truth to state. But it is the truth, and understanding it is essential to understanding both the legislative response that followed and the profound inequality that response has so far embodied.

The moment that crystallized the issue for the entertainment industry the moment that transformed AI deepfakes from a peripheral tech concern into a live political emergency came in May 2024, when Scarlett Johansson publicly confronted OpenAI.

Scarlett Johansson and the Voice That Was Not Hers

Scarlett Johansson is one of the most recognizable voices in American popular culture. Her performance as the AI character Samantha in the 2013 film Her warm, intimate, slightly husky, unmistakably human in its texture created an indelible association between her voice and the concept of AI-human connection.

In September 2023, Sam Altman, the CEO of OpenAI, contacted Johansson's agent to ask whether she would be willing to provide her voice for a new AI assistant that OpenAI was developing. Johansson declined. In March 2024, Altman reached out again. Johansson again declined, citing personal reasons. She was still in the process of considering the proposal when, in May 2024, OpenAI launched its new voice assistant GPT-4o with a default voice called Sky.

Sky sounded like Scarlett Johansson. Not approximately like her. Not reminiscent of her. Like her. The similarity was so striking that when OpenAI's own CEO posted on social media about the launch, he used a single word: Her. A direct reference to the film whose central character Johansson had voiced.

Johansson retained legal counsel. She contacted OpenAI directly. She released a public statement describing her shock and anger, noting that she had twice declined the company's request and that the voice had been created anyway or so it appeared. OpenAI maintained that Sky had been created independently, using a different voice actor, and was not derived from Johansson's voice. But the company did withdraw the Sky voice from public use, and the episode revealed something important: even the world's most famous actress, who had specifically and explicitly declined to license her voice, had no legal recourse that was clear, fast, or effective. There was no federal law making it illegal to create a voice that sounded like hers. There was no easy injunction available. Her only tools were public pressure and litigation and the public pressure worked primarily because she was Scarlett Johansson.

> **SCARLETT JOHANSSON:** She publicly stated that she had twice declined OpenAI's requests to use her voice, yet was left stunned and angered when the company launched an AI voice that was described by many as indistinguishable from hers. OpenAI withdrew the voice following her legal challenge.

Tom Hanks: The Fake Ad Problem

Tom Hanks is one of the most trusted public figures in America a two-time Academy Award winner whose persona of decency and reliability is among the most valuable in the entertainment industry. Which is exactly why, in 2023, his face and voice began appearing in advertisements for dental plans and other products that he had never endorsed and would never endorse.

Deepfake videos featuring a convincing Tom Hanks selling dental insurance began circulating on social media platforms, generating millions of views before they were reported and removed. In October 2023, Hanks was forced to post a warning on his own Instagram account, telling his followers that a video circulating of him promoting a dental plan was a fake that he had nothing to do with it, had never approved it, and wanted his followers to know so they would not be deceived.

The experience of being used as an unwitting shill having his face and voice deployed to sell products without his knowledge or consent, while being powerless to stop it beyond posting personal disclaimers on social media galvanized Hanks as a public advocate for legislative protection. He has spoken publicly and repeatedly about the threat, and his advocacy contributed to the political momentum behind the legislative response.

But the deeper lesson of the Hanks deepfake ads is not about Tom Hanks. It is about what happens when the most trusted faces in American culture can no longer be trusted. When a video of Tom Hanks recommending a product cannot be taken at face value, the entire architecture of visual trust begins to crumble. If not Hanks, then who? If not a face that has been associated with honesty and decency in the public mind for decades, then whose endorsement means anything? The deepfake crisis does not just steal identities. It corrodes the very concept of visual authenticity.

Morgan Freeman, Taylor Swift, and the Intimacy of the Stolen Voice

Morgan Freeman's voice one of the most distinctively recognizable in the English-speaking world began appearing in AI-generated audio content without his consent or knowledge, used for everything from promotional material to entertainment content that bore no relationship to anything he had actually said or would say.

The case of Taylor Swift represented a different and darker dimension of the deepfake crisis. In January 2024, sexually explicit AI-generated images depicting Swift were posted on X (formerly Twitter) and spread virally, reaching tens of millions of views before the platform took action to remove them. The images were not real photographs. They were entirely AI-generated. But they depicted Swift's face in explicit sexual situations that she had never been part of, had never consented to, and had no ability to prevent.

Swift's experience was, in one sense, the most famous instance of a category of abuse that affects millions of ordinary women and girls the creation and distribution of non-consensual intimate imagery using AI tools. But the scale of the public response to the Swift case the outrage, the congressional attention, the media coverage dwarfed the response to the same abuse perpetrated against people who were not globally famous. The technology is the same. The harm is the same. The difference is the victim's fame.

The TAKE IT DOWN Act federal legislation prohibiting the publication of non-consensual intimate images including AI-generated deepfakes was signed into law in May 2025. It passed with the support of both parties, moving with unusual speed through a Congress that routinely struggles to pass basic legislation. The groundwork for that speed was laid, in no small part, by the public fury over what happened to Taylor Swift.

> **TAKE IT DOWN ACT:** Signed into law May 2025, this legislation gives people the ability to request removal of non-consensual synthetic images from online platforms within 48 hours. SAG-AFTRA championed the bill since 2024. The law was accelerated by the public response to

deepfake images of Taylor Swift that went viral in January 2024.

The Hollywood Strike: When the Powerful Organized

The most dramatic and consequential organized response to the AI impersonation threat came from the entertainment industry and it offers a masterclass in how an organized, resourced, politically connected group can translate outrage into enforceable protection with remarkable efficiency.

In July 2023, the Screen Actors Guild – American Federation of Television and Radio Artists declared a strike against the Alliance of Motion Picture and Television Producers. One hundred sixty thousand actors, singers, and performers walked off the job. Production across Hollywood ground to a halt. The economic pressure was immediate and enormous.

The strike lasted 118 days. It was the longest SAG-AFTRA strike in history. And while it involved multiple grievances streaming residuals, compensation structures the issue of artificial intelligence was, from the beginning, at its beating heart.

The Body Scan That Lasts Eternity

The specific AI provision that had inflamed actors most was not hypothetical. It was already in practice. Studios had begun requiring background performers the actors who populate crowd scenes, who stand in the background of restaurants and airports, who give films their sense of inhabited reality to submit to full-body digital scans as part of their employment. The compensation was a single day's pay. The terms, in some contracts, permitted the studio to use the digital replica 'for the rest of eternity.'

Read that language again. For the rest of eternity. One day's pay in exchange for the permanent, unlimited use of your digital likeness your face, your body, your physical identity for any purpose the studio chose, forever, without additional compensation, without your ongoing consent, without any ability to opt out of future uses you had never imagined or agreed to.

This was not a worst-case interpretation of ambiguous contract language. This was what the studios were actually requiring. Background actors people who are, by definition, not powerful enough to negotiate individual contracts, who work for day rates and do not have agents skilled in intellectual property law were being asked to sign over their physical identity in perpetuity for the price of a shift.

When this practice became widely known when actors described it to journalists and on social media, when the full implications of 'for the rest of eternity' were spelled out in plain English the public response was immediate and visceral. Because it was not difficult to understand. You did not need to be a lawyer or a technologist to grasp what it meant. A studio wanted to own your face forever, for one day's pay.

SAG-AFTRA made the digital likeness provisions a central demand of the strike, and the union's leadership framed the issue with moral clarity that cut through the complexity of contract negotiations. Fran Drescher, the union's president, stated that AI posed an existential threat not only to SAG-AFTRA members but to civil discourse, student health and welfare, and democracy and national security. Duncan Crabtree-Ireland, the union's chief negotiator, called the NO FAKES Act the federal legislation that the union had drafted and championed an essential piece of legislation to restore individual rights.

The November 2023 Agreement: What the Powerful Won

When the strike ended in November 2023, SAG-AFTRA had secured a contract that represented, in the union's own assessment, historic protections for its members in the age of artificial intelligence. The key provisions are worth examining in detail, because they define the standard of protection that organized, powerful people were able to demand and that ordinary people have not yet been able to obtain.

Consent. No actor's likeness or performance may be replicated digitally without prior written approval. Not implicit approval buried in a general employment contract. Explicit, specific, written consent for each use.

Compensation. If a digital replica is used, the actor must be paid fairly for that use whether through upfront fees or residuals. The use of a digital likeness is treated as labor, compensated as labor.

Control. Actors retain the right to decline digital replication entirely. They retain the right to limit how their image is used. They cannot be forced, as a condition of employment, to surrender their digital likeness.

Specificity. The intended uses of any digital replica must be described in specific, comprehensible terms before consent is given. 'For the rest of eternity' is not an acceptable description of intended use.

These are reasonable protections. They are, in fact, the minimum reasonable protections for any human being whose physical identity might be digitally replicated. The principle that your face and voice belong to you that their use requires your specific consent and fair compensation is not radical or controversial. It is the obvious baseline that any ethical framework would derive from first principles.

But these protections apply to SAG-AFTRA members. They apply to the 160,000 performers with union cards. They do not apply to the 330 million Americans who are not SAG-AFTRA members. They do not apply to Sharon Brightwell's daughter, whose voice was cloned from Instagram. They do not apply to the elderly couple whose life savings were stolen by someone impersonating their grandchild. They do not apply to you.

SAG-AFTRA PROTECTION: Union members now have legally binding rights to consent, compensation, and control over digital replicas of their face and voice. These protections were secured through 118 days of industrial action backed by 160,000 organized workers. They do not extend to non-union members or ordinary citizens.

California, New York, and Tennessee: State-Level Protection for the Famous

Following the SAG-AFTRA agreement, state legislatures particularly in California, New York, and Tennessee moved to codify some of these protections into law, extending them beyond union membership.

California's AB 2602, signed by Governor Gavin Newsom in September 2024 and effective January 1, 2025, requires a performer's contractual consent and proper representation before using a digital replica of their voice or likeness. The law protects artists from being bound to contracts that allow the use of their digital likeness without full, informed, specific consent and applies both to future contracts and to existing contracts that would generate future digital replica uses.

New York enacted substantially identical legislation in December 2024. Tennessee, moved by the particular relevance to the country music industry that dominates its cultural economy, passed the ELVIS Act Ensuring Likeness Voice and Image Security in March 2024, making it the first state to explicitly protect individual voice and likeness against unauthorized AI use.

These are genuine legislative achievements. They represent real protections for real people. But their scope reveals as much as their substance. California's law is primarily aimed at protecting performers in the entertainment industry. New York's mirrors it. Tennessee's was explicitly framed around the concerns of musicians. The legislative momentum that produced these laws was generated by a lobbying apparatus representing the entertainment industry by SAG-AFTRA and the Recording Industry Association of America and the Motion Picture Association not by consumer advocacy groups representing ordinary people.

The laws protect performers because performers organized and demanded protection. The lesson is clear and deliberate: protection in the AI age, as in every age, goes to those with the power to demand it.

The NO FAKES Act: Federal Protection Drafted by Those Who Have the Most to Lose

The NO FAKES Act Nurture Originals, Foster Art, and Keep Entertainment Safe is the most significant proposed federal legislation addressing AI deepfakes and digital likeness rights. It was introduced in the Senate in July 2024 by a bipartisan coalition of senators. It was drafted in collaboration with SAG-AFTRA, the Recording Industry Association of America, and the Motion Picture Association. Its list of supporters reads like a roster of the most powerful interests in the entertainment and technology industries.

The bill would prohibit creating or distributing a digital replica of a living or deceased person without their consent. It would hold individuals and companies liable for damages. It would impose obligations on platforms that knowingly host unauthorized deepfakes. It would create, for the first time, a federal intellectual property right in one's own voice and visual likeness.

These are meaningful protections. If enacted, the NO FAKES Act would represent a significant step toward a legal framework that treats every person's identity as their own property. The bill explicitly extends beyond performers to 'individuals' meaning, in principle, ordinary people as well as celebrities.

But the political economy of how the bill was drafted and advanced tells a revealing story. SAG-AFTRA's national executive director called the bill 'essential legislation to restore individual rights' but the union's primary concern was its 160,000 members. The Motion Picture Association's support was predicated on protecting its member studios' legitimate uses of AI while constraining unauthorized ones. The Recording Industry Association's interest was in protecting the economic value of established artists' voices. OpenAI and IBM technology companies with significant stakes in how AI regulation develops were among the bill's supporters.

The ordinary grandmother who loses her savings to a deepfake grandchild scam has no seat at this table. No lobbyist. No political action committee. No strike fund. No institutional voice. She is the nominal beneficiary of legislation whose real driving forces are industry interests that sometimes align with her protection and sometimes do not.

The Inequality of Protection: Who Gets Covered and Who Does Not

The current landscape of AI deepfake protection in the United States reflects a familiar pattern in American law and policy: protections emerge first for the powerful, trickle slowly toward the ordinary, and often stop well short of the most vulnerable.

Consider who is currently protected and who is not. A SAG-AFTRA member whose likeness is used without consent has a specific, enforceable contractual right, backed by union legal resources, to seek remedy. A California resident who is a performer has statutory rights under AB 2602. A Tennessee musician has rights under the ELVIS Act. A person whose non-consensual intimate image is distributed online has rights under the TAKE IT DOWN Act though enforcement depends on platform compliance with removal requests.

Now consider who is not protected. An elderly retiree whose voice is cloned from a voicemail and used to scam their family has no specific federal cause of action for the voice cloning itself. A small business owner whose face is put in a fake video endorsing a competitor's product has limited and expensive legal remedies. A teenager whose image is used in AI-generated content without consent has protections only in states that have passed specific legislation. A worker whose identity is stolen and used to apply for jobs and bank accounts a growing form of synthetic identity fraud has fraud remedies but no specific AI identity protection.

The gap between who is protected and who needs protection is not a gap of deserving. It is a gap of power. The people who have protection organized, lobbied, and spent money to get it. The people who need protection have not yet been organized, and the machinery of American legislative politics has not yet felt the pressure necessary to act on their behalf.

> **THE GAP:** SAG-AFTRA secured specific, contractual, legally enforceable protections for 160,000 union members in November 2023. There are 335 million Americans. The other 334.84 million are protected by a patchwork of general fraud law, state statutes, and proposed federal legislation that has not yet been enacted while deepfake fraud accelerates at 700% annually.

The Human Cost: Real Victims, Real Devastation

Numbers and legislative analysis can create a false distance from the human reality of what deepfake fraud actually does to people. Let us close that distance. The stories that follow are documented cases real people, real losses, real devastation that represent not exceptions but the accelerating norm.

Steve Beauchamp: The Man Who Believed Elon Musk

Steve Beauchamp is 82 years old. He spent his working life building something accumulating savings that represented decades of labor and sacrifice, money that was supposed to carry him through his final years with security and dignity.

Steve Beauchamp no longer has that money.

Beauchamp came across a video online of Elon Musk one of the most recognizable faces in the world, a man whose technological credibility and financial success are known to virtually every American adult. In the video, Musk was promoting a cryptocurrency investment opportunity that promised extraordinary returns. The video was convincing. Musk's face moved naturally. His voice matched. His mannerisms the slight upward glance, the measured cadence of explanation were authentic.

Everything about it was real, except that none of it was real.

Beauchamp invested. And then he invested more. Over the course of several weeks, convinced that a man he recognized, whose face he could see, whose voice he could hear, was personally recommending an investment, Beauchamp transferred six hundred and ninety thousand dollars his retirement fund, the savings of a lifetime to fraudsters.

When the deception was uncovered, his response to reporters captured something that no policy document or legislative brief ever quite conveys: 'I mean, the picture of him it was him.'

Of course it was. That is the point. That is what makes this crime different from every other form of fraud in human history. The psychological safeguards that human beings use to protect themselves from deception the ability to recognize a face, to identify a voice, to read the physical cues of authentic human presence all failed, because they were defeated by a technology specifically engineered to defeat them.

Steve Beauchamp lost his retirement to a technology he did not know existed, deployed against a vulnerability he had no reason to know he had, in pursuit of a crime that carries minimal legal risk for its perpetrators and has no specific federal law prohibiting the tool used to commit it.

Sharon Brightwell: The Daughter's Voice

We opened this chapter with Sharon Brightwell's story. But let us return to it with more detail, because the mechanics of what happened to her illuminate the machinery of the threat with a clarity that statistics cannot provide.

Sharon's daughter had a public Instagram account. Her daughter posted videos occasionally casual, personal content of the kind that hundreds of millions of people post every day without any thought for who might be recording it, analyzing it, or feeding it into an AI system. From these videos, someone extracted audio. From that audio less than a minute of someone's voice, speaking casually about ordinary things a voice clone was generated. A digital replica, indistinguishable to a mother's ear, of her daughter's voice in distress.

The total cost of the tools required: a few dollars in subscription fees, available to anyone. The technical skill required: none that a moderately motivated person could not acquire in an afternoon. The emotional devastation to Sharon Brightwell: immeasurable. The legal recourse available to her: a fraud report to the FTC and local law enforcement, which are overwhelmed by the scale of such cases and have limited capacity to pursue individual losses of fifteen thousand dollars.

Sharon Brightwell is not a victim of unusual circumstances. She is a victim of ordinary circumstances in 2025. The only thing that distinguishes her case from thousands of others that occur monthly is that her story was documented and reported. The vast majority are not.

The $25 Million Video Call

In February 2024, an employee of Arup a British engineering company with global operations received an invitation to a video conference call. The call appeared to include the company's Chief Financial Officer and several other senior executives. The faces were familiar. The voices matched. The conversation was normal a discussion of business matters, a request for a financial transaction, the ordinary choreography of corporate life.

The employee authorized fifteen wire transfers totaling approximately $25 million to Hong Kong bank accounts.

Every person on the video call except the employee was a deepfake. The CFO was a deepfake. The executives were deepfakes. The entire meeting was a synthetic construction, generated by AI systems using available footage and audio of the company's leadership. No one on that call was who they appeared to be. And the employee, using every visual and auditory cue available to human perception, could not tell.

Arup's global chief information officer described in the aftermath a sharp rise in the number and sophistication of such attacks. The $25 million loss the largest single documented deepfake fraud at the time was not, in his assessment, an anomaly. It was a preview.

It was also a demonstration that the deepfake threat is not limited to the elderly and the vulnerable. A professional employee at a sophisticated global corporation, presumably with security awareness training, with access to institutional resources, in a business environment with established procedures could not detect a deepfake meeting that cost his company $25 million. The technology does not discriminate. It defeats trained professionals as efficiently as it defeats retired grandmothers.

The 82-Year-Old and the Grandchild in Jail

The grandparent scam in which fraudsters impersonate a grandchild claiming to be in legal trouble and needing immediate financial help predates AI. It is one of the oldest telephone fraud schemes in existence. For decades, it has worked through social engineering alone: an unfamiliar voice, a stress-filled story, the exploitation of a grandparent's love and protective instinct.

AI voice cloning has transformed this already devastating crime into something categorically more dangerous. The unfamiliar voice has been replaced by a familiar one the actual voice of the actual grandchild, replicated from social media audio with enough fidelity to pass the biological recognition test that every parent and grandparent carries.

An elderly couple documented by consumer protection organizations transferred their life savings after receiving a voice message from someone whose voice was their grandchild's voice the specific timbre, the specific accent, the specific emotional texture telling them their grandchild was in jail and needed bail money immediately. The urgency was real. The fear was real. The love was real. The voice was not.

The FBI's elder fraud data consistently shows that investment scams and impersonation scams are the most costly forms of elder fraud. AI voice cloning has turbo-charged the impersonation category by defeating the primary safeguard that protected against it the ability to recognize a family member's voice.

FBI 2023 ELDER FRAUD DATA: Elder fraud costs Americans over $3.4 billion annually, with impersonation scams among the most damaging categories. AI voice cloning has fundamentally changed the threat landscape for elderly victims by defeating the voice-recognition safeguard that was previously their most reliable protection.

The Non-Consensual Intimate Image Epidemic

Before we proceed, we must address a dimension of the AI impersonation crisis that is, in some ways, the most immediately harmful to the largest number of victims, and that is also the dimension that has received the least proportionate attention relative to its scale: the generation and distribution of non-consensual AI-generated intimate imagery.

This is not a comfortable topic. It is a topic that involves sexual abuse, psychological devastation, and the systematic targeting of women and girls by technology that has been deliberately designed to remove the barriers to creating explicit content depicting real people without their consent.

The scale is staggering. Non-consensual intimate images including AI-generated deepfakes accounted for 32% of all deepfake cases documented by researchers in 2024. The victims are overwhelmingly female. The perpetrators are predominantly male. The technology has created a new form of sexual abuse that can be perpetrated against any woman whose image appears online which is to say, virtually every woman alive.

From Famous to Ordinary: The Democratization of Abuse

When AI-generated intimate images of Taylor Swift went viral in January 2024, reaching an estimated 27 million views before X removed them, it produced the moment of public outrage that drove legislative action. The TAKE IT DOWN Act that resulted is a genuine step forward. Mandatory removal within 48 hours of reporting is meaningful.

But Taylor Swift's experience, as devastating as it was, differs from that of ordinary women in one critical dimension: she had the resources, the legal team, the public platform, and the cultural power to compel a response. When the same technology is used against a high school girl in a suburban town when her image is used to generate explicit content that is then shared among her classmates the machinery of response looks very different.

The TAKE IT DOWN Act requires platforms to remove content within 48 hours of a valid request. It does not prevent the content from being created. It does not make the creation of such content a federal crime with meaningful enforcement. It does not provide victims with any remedy beyond removal not damages, not criminal accountability for the perpetrators, not the ability to ensure that content removed from one platform does not immediately reappear on another.

And the removal requirement applies only to platforms that comply with it. Much of the most harmful non-consensual intimate content circulates not on major platforms but on dedicated websites, private messaging applications, and dark web forums that have no intention of complying with any removal requirement and face minimal realistic enforcement risk.

The Psychological Devastation

The harm of non-consensual intimate imagery AI-generated or otherwise is not primarily financial, though it can be. It is psychological, social, and existential. Victims describe the experience of discovering that explicit content depicting their likeness exists online as a violation that is difficult to distinguish, in its subjective experience, from physical assault. The sense of bodily violation, of having one's most intimate self deployed without consent for the gratification or entertainment of strangers, produces trauma responses that are well-documented and severe.

For young people for the teenagers who are among the most common targets of peer-generated non-consensual intimate imagery the effects can be catastrophic. Academic performance collapses. Social relationships fracture. Depression and anxiety follow predictably. In documented cases, victims have left schools, abandoned careers they had worked toward, and in the most tragic cases, taken their own lives.

There is no deepfake removal tool that undoes that damage. There is no law that can restore what is lost once a person has experienced the specific violation of seeing their face on a body they did not consent to display. The only meaningful protection is preventing the creation and distribution of such content in the first place which requires both technical measures and legal consequences serious enough to deter perpetrators.

The current legal framework, even with the TAKE IT DOWN Act, falls well short of that standard.

The Business of Stolen Identity: Corporate Deepfake Fraud

The devastation that deepfakes wreak on individuals is the most emotionally resonant dimension of the crisis. But the corporate and institutional dimensions of AI impersonation fraud deserve equal attention both because the financial scale is enormous and because the failure of businesses to protect themselves has direct consequences for ordinary employees and customers.

Business Email Compromise Goes Video

Business email compromise in which fraudsters impersonate executives or business partners via email to authorize fraudulent financial transfers has been one of the fastest-growing categories of cybercrime for years. In 2023, the FBI's Internet Crime Complaint Center reported that BEC losses exceeded $2.9 billion in the United States alone.

AI deepfakes have added a video and audio dimension to this already devastating threat. Whereas previously a business email compromise attack was limited by the fact that a CFO could be verified by phone call, deepfake technology has eliminated phone verification as a security measure. And as the Arup case demonstrated, video verification is also no longer reliable.

The attack surface has expanded to every meeting, every call, every video conference in which financial authorization or sensitive information transfer occurs. Every communication channel through which business decisions are made is now a potential vector for AI impersonation fraud.

A 2024 Deloitte survey found that more than one in four executives reported that their organization had experienced at least one deepfake incident, with 50% expecting attacks to increase in the following year. Average losses per deepfake-related corporate incident in 2024 were nearly $500,000, with large enterprise losses reaching as high as $680,000. These are not costs that corporations can absorb indefinitely, and they are costs that ultimately flow through to customers, shareholders, and employees.

Synthetic Identity Fraud: Becoming Someone Who Never Existed

A particularly sophisticated application of AI deepfake technology is synthetic identity fraud the creation of entirely fictitious human identities using AI-generated images, voices, and partial real data.

Synthetic identity fraud is not the same as identity theft. Identity theft involves stealing a real person's existing identity. Synthetic identity fraud involves creating a new person who does not exist combining a real Social Security number (often belonging to a child or elderly person with minimal credit history) with AI-generated photos, AI-generated voices, and fabricated documentation to create a convincing persona that can open bank accounts, apply for credit, take out loans, and conduct financial activity.

Synthetic identity document fraud surged 300% in the United States in the early months of 2025 alone, according to Sumsub. The AI tools required to create a convincing synthetic identity realistic photos from image generation models, convincing voice samples from voice cloning tools, plausible supporting documentation from document generation tools are all commercially available. The barrier is not technical sophistication. It is the will to use the tools for fraud.

Perhaps the most alarming documented case of synthetic identity fraud in recent memory was the discovery that North Korean IT workers were using AI-generated deepfake videos to successfully apply for remote employment positions at American technology companies passing video interviews, providing synthetic identity documentation, and gaining employment that provided both a salary and access to company systems. The same technology that is used to steal grandmothers' savings is being used by state intelligence actors to infiltrate American corporations.

The Ferrari Test and Real-Time Deepfakes

In 2024, an executive at Ferrari received a WhatsApp voice message that appeared to be from the company's CEO, discussing confidential business matters and requesting an unusual action. The executive alert, trained, operating in a security-conscious environment paused. Something was slightly off. The executive verified through an independent channel. It was a deepfake. The attack was thwarted.

But the 'Ferrari test' the standard of requiring executives to be alert enough to detect real-time AI impersonation of their own CEO through voice messaging is not a realistic security standard for most employees, in most organizations, under most conditions. Human beings are not designed to function in a permanent state of verification skepticism toward the people they work with. The cognitive overhead of treating every communication as potentially fraudulent is incompatible with functional organizational life.

The deeper problem is that as real-time video deepfakes become more sophisticated and more accessible, the standard of 'something seemed slightly off' becomes less and less reliable as a detection mechanism. Investigators at 404 Media documented that scammers in 2024 were already using real-time face-replacement tools during live video calls with sufficient fidelity that they were passing visual verification with trained professionals.

The future of this technology is not 'slightly off.' The future is indistinguishable. And the organizations, institutions, and individuals who are planning their defenses around the assumption that careful attention will reveal the fake are planning for a threat that is already being superseded.

The Inequality at the Heart of the Crisis

We have now covered the technology, the victims, the powerful, and the legislation. Let us bring the central argument of this chapter into full focus, because it deserves to be stated clearly and without diplomatic softening.

The people who built the AI impersonation threat the technology companies made billions in venture capital and revenue while doing it. The people who first felt the threat most acutely famous actors, celebrated musicians, globally recognized billionaires had the resources, the organization, and the political connections to respond quickly and effectively. The laws and contracts they secured protect them and, to a limited extent, other members of their professional communities.

The people who are now bearing the greatest burden of the AI impersonation crisis elderly retirees losing their life savings, young women whose intimate images are distributed without consent, ordinary employees tricked into fraudulent wire transfers, families deceived by cloned voices of loved ones have received, by comparison, almost nothing.

This is not an accident. It is the predictable output of a political and regulatory system in which protection goes to those with the power to demand it. The entertainment industry demanded protection and got it. Ordinary people have not yet demanded protection with sufficient organization, volume, and political force to get it. That is the situation. It is a solvable problem. But solving it requires understanding it clearly.

The Outrage That Should Be Yours

Consider the following comparison. When Scarlett Johansson discovered that a company had created an AI voice that sounded like hers after she had specifically declined to license her voice the resulting public controversy caused that company to withdraw the voice, compelled the company's CEO to address the matter personally, and contributed to congressional momentum for federal legislation. The entire process, from discovery to partial remedy, took weeks.

When Sharon Brightwell's daughter's voice was cloned from Instagram and used to steal fifteen thousand dollars from her mother without consent, without compensation, without any legal protection for the voice's owner the remedy available was a fraud report to the FTC that is one of millions filed annually, processed by an agency that lacks the resources to individually investigate the vast majority of them.

The voices of both women were taken without consent. The harm in both cases was real. The response was entirely different, because one of the women is a globally famous actress with legal resources, media access, and institutional power, and the other is an ordinary person.

This is the outrage that should be yours. Not directed at Scarlett Johansson she did nothing wrong by seeking protection for herself. Not directed at SAG-AFTRA they did their job, which was to protect their members. The outrage belongs directed at the system that treats identity protection as a luxury good available to the powerful rather than a universal right belonging to every human being.

Tom Hanks needed a law to protect his face. You need the same law. Taylor Swift needed legislation to protect her against non-consensual intimate imagery. So does every woman and girl in the country. Steve Beauchamp needed protection against deepfake financial fraud. So does every elderly person with savings and a social media presence.

The protection that the famous and powerful obtained through organization and political will is not more than what ordinary people deserve. It is less. The famous have legal teams and public platforms and institutional allies. The ordinary have none of these things. They need the law more, not less. And right now, they have it less.

The Three-Class System of AI Impersonation Protection

What has emerged in the United States is effectively a three-class system of protection against AI impersonation.

The first class consists of organized, institutionally represented workers in the creative industries SAG-AFTRA members, recording artists signed to major labels, others covered by specific collective bargaining agreements or state laws. They have the highest level of protection: specific contractual rights, union legal resources, state statutory protections in California, New York, and Tennessee. If their likeness is used without consent, they have a clear legal path to remedy.

The second class consists of individuals in states that have passed broader consumer protection legislation covering digital likeness rights, or those with resources to pursue civil litigation for violations. They have moderate protection: state statutory rights in a limited number of jurisdictions, general fraud remedies, and the partial protection of the TAKE IT DOWN Act for intimate imagery. Their protection is real but incomplete, geographically uneven, and expensive to enforce.

The third class the vast majority of Americans consists of everyone else. People who live in states without specific digital likeness protection. People who cannot afford to litigate. People whose voices and faces are scraped from social media daily by systems they are not aware of, processed by AI models they did not consent to train, and potentially used in ways that range from annoying to devastating. Their primary protection is general fraud law the same law that existed before AI was invented, applied to a threat for which it was not designed and for which it is often inadequate.

This is the landscape as of 2025. It is not stable. It is not acceptable. And it will not change unless ordinary people understand it clearly enough to demand change.

The Technology Accelerating Ahead of the Law

Every legislative and contractual protection described in this chapter was written in response to threats that existed when the writing happened. The technology has not stood still. It is advancing faster than any legislative process can follow, and the capabilities that are emerging in the near term will make current deepfakes look primitive.

Emotional AI: Replicating the Soul

Current deepfake technology replicates appearance and voice with high fidelity. The next generation will replicate emotional response and personality with comparable fidelity. AI systems are being trained not just to reproduce how a person looks and sounds but to model how they think, how they respond emotionally, how they behave under different conditions.

An emotional deepfake will not just say words in your voice. It will respond to questions in ways that reflect your opinions, your values, your characteristic emotional responses, your relationship with the person it is deceiving. When your elderly parent receives a call that not only sounds like you but responds to their questions as you would, acknowledges inside family jokes as you would, expresses concern for the things you actually care about the deepfake will have crossed a threshold from audio impersonation to identity impersonation.

The psychological manipulation potential of emotional AI is without precedent in the history of fraud. Every scam that exploits human relationships every fraudulent appeal to love, to family loyalty, to trust built over decades will be amplified by a technology that can simulate the specific person who occupies those relationships in any given target's life.

Autonomous AI Scam Agents: Scale Without Limit

Current deepfake scams require human operators. A scammer must conduct the conversation, respond in real time, maintain the deception through the interaction. This human requirement limits scale. There are only so many scammers, operating in only so many locations, capable of conducting only so many simultaneous deceptions.

The next generation of AI fraud will not be limited by human operators. Autonomous AI agents systems that can conduct complete, contextually appropriate, emotionally sophisticated conversations without human involvement will be able to run thousands of deepfake scam interactions simultaneously, each customized to its target based on data harvested from social media, each maintaining a convincing persona for the duration of whatever interaction is required to complete the fraud.

One person with a laptop and the right tools will be able to run a fraud operation targeting thousands of victims simultaneously. The $200 million in deepfake losses recorded in the first quarter of 2025 will look modest by comparison with what autonomous AI fraud agents will be capable of in the next five years.

The Biometric Authentication Crisis

Modern security systems are increasingly built around biometric authentication the use of your face, your voice, your fingerprint, your iris pattern as the key to your digital identity. Banks use voice verification for account access. Government services use facial recognition for identity verification. Employers use voice and face authentication for remote access to sensitive systems.

All of these authentication systems assume that your biometric signature your face, your voice cannot be replicated by someone who does not have physical access to you. AI deepfakes have invalidated that assumption. When a voice clone can pass a voice authentication system and a real-time video deepfake can pass a facial recognition system, biometric authentication is no longer authentication. It is a checkpoint with a forgeable pass.

The financial industry has been among the first to confront this reality. Deloitte and other financial industry analysts have identified biometric spoofing as one of the central challenges for the sector's security infrastructure. The authentication systems that were designed to replace passwords because passwords could be stolen are now themselves vulnerable to a different kind of theft: the theft of the biometric identities they were supposed to protect.

The Practical Defense: What You Can and Must Do Now

This chapter has traced a crisis from technology to victimization to legislative response to the deep inequality of that response. It would be irresponsible to leave the reader here with fear and no agency. So let us be direct and specific about what individuals can do to protect themselves in the current environment, acknowledging clearly that individual action is necessary but not sufficient, and that systemic change requires political action.

Your Voice Is Already Out There Minimize New Exposure

Every video you have posted on social media, every voice message you have sent, every podcast appearance, every YouTube video, every TikTok all of it is potential training data for a voice clone of you. You cannot take back what is already public. You can reduce future exposure.

Consider adjusting the privacy settings on your social media accounts to limit who can access your content. Consider whether every video you post serves a purpose that outweighs the incremental risk of providing more voice sample data. Consider that the person who needs to worry most is not you it is your elderly parents, whose voices are relatively unfamiliar to the scammers who might target them, and your children, whose voices can be cloned to target you.

This is not a counsel of digital silence. It is a counsel of awareness. Know what you are sharing and with whom, and understand that your voice and your face are now assets with security implications as well as social ones.

Establish Family Verification Protocols

The most effective defense against AI voice clone family scams is a protocol that your family agrees on before an emergency. A family code word a specific word or phrase that only family members know, which must be provided at the beginning of any call requesting money or immediate action is a simple, low-tech, highly effective countermeasure.

Establish it now, before you need it. Make sure every member of your family especially elderly parents and grandparents knows the code word and understands why it exists. Make it a rule: no money, no action, without the code word. Period.

Supplement this with a rule about verification: if you receive a call from a family member claiming to be in trouble, hang up and call them back directly on a number you have previously used. Do not call back a number provided during the emergency call. Call the number in your contacts. If they answer, the emergency call was a scam. If they do not answer, leave a message and try alternative contacts before taking any financial action.

Treat Video Calls as Unverified Without Second-Channel Confirmation

For any financial transaction, sensitive information transfer, or significant decision prompted by a video or audio call, require second-channel verification before acting. Call the person back on a different number. Send an email to a known address. Check with a colleague through a separate communication channel. Do not let the urgency of the call which is often manufactured specifically to prevent this kind of verification override the protocol.

This applies in personal life and professional life. The Arup employee who authorized $25 million in wire transfers from a video call with deepfake executives had no protocol requiring second-channel verification for unusual financial requests. That protocol, applied consistently, would have prevented the loss.

Organizations should formalize this into policy: multi-person authorization for financial transactions above a threshold, out-of-band verification requirements for unusual requests, and explicit guidance that unusual urgency is itself a red flag warranting additional verification rather than faster action.

Be Specifically Skeptical of Investment Opportunities Promoted by Famous People

The deepfake Elon Musk phenomenon convincing videos of the world's most famous entrepreneur promoting cryptocurrency investment opportunities has cost ordinary people billions of dollars globally. This specific attack vector has a simple defense: assume any investment opportunity promoted in a video featuring a famous person is fraudulent until proven otherwise through multiple independent verification steps.

No legitimate investment opportunity requires you to act immediately based on a video you encountered online. No legitimate financial opportunity will be withdrawn if you take 48 hours to verify it through official channels. Urgency is a manufactured scarcity. It is a manipulation technique. Real investments do not disappear if you verify them. Scams do.

Know Your Rights and Document Everything

If you are the victim of AI deepfake fraud or non-consensual use of your likeness, document everything immediately. Screenshots, recordings, URLs, timestamps, transaction records preserve all of it before anything can be taken down or deleted.

Report to the FTC at reportfraud.ftc.gov. Report to the FBI's Internet Crime Complaint Center at ic3.gov. Report to local law enforcement. Report to whatever platform the deepfake appeared on.

None of these actions guarantees a remedy the reality is that individual cases are often too small and too technically complex for overwhelmed law enforcement agencies to investigate individually. But reporting creates the data record that drives policy attention, and in some cases, particularly for larger financial losses, triggers investigation.

In states with specific digital likeness protection laws California, New York, Tennessee consult an attorney about your rights. In states without such laws, general fraud and tort remedies may still be available, and an attorney with cybercrime experience can assess whether civil litigation is viable.

The Political Fight: Demanding Equal Protection

Individual defensive measures protect individuals. They do not change the system that has generated the crisis. Changing the system requires political action and understanding what political action is available.

The NO FAKES Act Must Pass For Everyone

The NO FAKES Act, if enacted as written, would create a federal right for every individual not just performers, not just union members to control the use of their voice and visual likeness in AI-generated content. This is the baseline protection that the current legal landscape lacks.

The bill has bipartisan support. It has industry support from both the entertainment sector and major technology companies. Its passage has been delayed by the ordinary entropy of American legislative politics competing priorities, procedural obstacles, election cycles that reset the clock on legislative momentum.

Contacting your senators and representatives in support of the NO FAKES Act is a concrete, actionable step. Congressional offices count calls and emails. Constituent pressure moves legislation. The bill's momentum depends on ordinary people communicating to their elected representatives that they care about this issue that it affects them personally, that they want it passed, and that they will remember how their representatives voted.

Demand Consumer-Level Enforcement, Not Just Industry Protection

As the AI identity protection legislative landscape develops, pay close attention to who benefits. Legislation drafted primarily by industry lobbyists tends to protect industry interests first and consumer interests second. Legislation that includes strong civil remedies for individual victims actual damages, statutory damages, attorney fee shifting that makes it financially viable for individuals to bring claims without deep pockets is categorically more useful to ordinary people than legislation that creates rights without remedies.

The difference between a law that says 'your likeness cannot be used without consent' and a law that says 'your likeness cannot be used without consent, and if it is, you can recover meaningful damages without spending $50,000 in legal fees to do so' is enormous in practical terms. Demand the latter. The former protects the rights of people who can afford to enforce them. The latter protects everyone.

The Equality Argument: Making the Case That Resonates

The most politically effective argument for comprehensive AI identity protection is not a technical argument about deepfake capabilities. It is an equality argument.

Tom Hanks has protection. Scarlett Johansson has protection. Taylor Swift triggered legislation that protects her and, to some extent, everyone else. The protection that famous and powerful people secured for themselves is not more than what ordinary people deserve it is the minimum standard of dignity and autonomy that any free society should guarantee to all its members.

If it is wrong for OpenAI to clone a famous actress's voice without her consent, it is equally wrong to clone a Florida mother's daughter's voice for a fifteen-thousand-dollar theft. If it is wrong for a studio to own a famous actor's digital likeness in perpetuity for one day's pay, it is equally wrong for a technology company to scrape your face from your Instagram and use it to train AI models without your consent or compensation. If it is wrong for non-consensual intimate images of a global superstar to be distributed online, it is equally wrong when it happens to your daughter.

The principle is the same. Your identity belongs to you. It is not a data point. It is not a training sample. It is not a resource for extraction. It is you your face, your voice, your presence in the world and it cannot be taken, replicated, or deployed without your specific, informed, freely given consent.

That is not a radical position. It is the most basic claim that any person can make about their own existence. And right now, in the United States of America in 2025, it is not fully protected by law.

The famous got their protection. Now it is time for the rest of us to get ours.

* * *

END OF CHAPTER THREE

Continue to Chapter Four: Your Money, Their Control

CHAPTER FOUR

Your Money, Their Control:

The Cashless Society, the Ownership Crisis, and the Slow Erasure of Ordinary Wealth

---

"Welcome to 2030. I don't own anything. I don't own a car. I don't own a house. I don't own any appliances or any clothes." — Danish politician Ida Auken, World Economic Forum essay, 2016

Eight Words That Captured the World's Attention

In November 2016, the World Economic Forum published a short video on its social media channels. The video was titled '8 Predictions for the World in 2030.' It was produced as a thought-provoking look at how dramatically life might change in the coming decade and a half. One of its eight predictions, drawn from an essay by Danish member of parliament Ida Auken, captured the world's imagination in a way that its creators probably did not anticipate.

The prediction was this: 'You'll own nothing, and you'll be happy.'

The full context of Auken's essay described an imagined city in 2030 where everything — cars, apartments, appliances, clothes — is rented or shared through digital platforms. Products are services. Access replaces ownership. The circular economy has eliminated waste and consumption. The narrator describes the arrangement with cheerful satisfaction.

The essay was speculative fiction, framed as a thought experiment rather than a policy roadmap. The WEF is an influential forum, not a government. It issues no laws, commands no armies, controls no budgets. Auken herself later clarified that the piece was meant to spark debate, not prescribe outcomes. Fact-checkers were quick to point out that the 'you will own nothing' framing had been stripped from its nuanced original context and transformed into a sinister slogan.

All of that is true. And none of it matters.

What matters is that the phrase landed with the force of recognition — not because it described a WEF conspiracy, but because it described something that millions of ordinary people were already living. The sentence resonated not as a revelation of secret plans but as a diagnosis of an observable reality that was already unfolding around them.

Homeownership rates declining for younger generations. Physical media replaced by subscriptions. Cars replaced by ride-shares. Retirement savings replaced by debt. The sense that the financial foundations that previous generations had built — the house, the car, the land, the savings — were becoming harder to reach, more precarious once obtained, and more likely to end up, ultimately, in the hands of banks, corporations, and landlords rather than children and grandchildren.

The WEF did not cause any of this. The phrase did not invent the problem. But it gave a name to something real. And this chapter is about that real thing — the convergence of financial surveillance, digital currency, institutional ownership, declining generational wealth, and the political economy of debt that is quietly rewriting the terms on which ordinary people participate in the economy.

This is not a chapter about conspiracy. It is a chapter about consequences — the documented, measurable, ongoing consequences of financial system changes that are largely invisible to the people most affected by them until it is too late.

The Ownership Society That Wasn't

To understand what is being lost, we must first understand what was there to lose. The concept of ordinary wealth — wealth accessible to working and middle-class families, not just the rich — has a specific history in the United States, and that history is built on a handful of pillars that previous generations took for granted as the basic architecture of a stable life.

The first pillar was the house. For most of American history, homeownership was the primary mechanism through which ordinary families built wealth. A home was not just a place to live. It was a store of value that appreciated over time, a hedge against inflation, a source of equity that could be borrowed against in emergencies, and an asset that could be transferred to children and grandchildren. The generational transmission of home equity — the house that parents paid off that their children inherited or received as a financial foundation — was one of the primary mechanisms through which economic advantage was passed from one generation to the next in working-class and middle-class families.

The second pillar was savings. Defined-benefit pension plans — in which employers bore the investment risk and guaranteed a specific monthly income in retirement — provided working people with financial security in old age that did not depend on their own investment acumen or the vagaries of financial markets. The shift from defined-benefit to defined-contribution plans over the past four decades has transferred the risk of retirement security entirely from employers to workers, converting what was a guaranteed foundation into a gamble on market performance.

The third pillar was tangible assets. Land, gold, physical goods — things that could be held, stored, transferred, and inherited without intermediaries. Tangible assets have historically been the refuge of ordinary wealth precisely because they are difficult to confiscate, difficult to inflate away, and impossible to freeze or cancel. A farmer's land and a craftsman's tools were wealth in a form that no bank run, no currency crisis, and no government policy could easily erase.

All three of these pillars are under sustained pressure in the contemporary economy. This chapter examines each, the forces acting upon it, and the digital financial architecture — including but not limited to central bank digital currencies — that could accelerate the erosion further.

The Housing Trap: When Shelter Becomes a Financial Product

The American Dream has always contained, at its center, a specific image: the owned home. Not rented. Owned. The deed in your name. The equity building month by month with every mortgage payment. The garden you can plant whatever you want in. The walls you can paint whatever color you choose. The place that belongs to you, that no landlord can sell from under you, that you can leave to your children.

That dream is becoming increasingly inaccessible for large portions of the American population, and the reasons are multiple, complicated, and hotly debated. But some of the documented forces are worth examining closely.

The Affordability Collapse

By virtually every metric, housing affordability in the United States has deteriorated to historic levels over the past decade. The National Association of Realtors' Housing Affordability Index — which measures whether a family earning the median income can qualify for a mortgage on a median-priced home — fell to its lowest recorded level in 2023, lower even than the peak of the 2006 housing bubble. The typical home now costs more than five times the median annual household income in the United States, compared to approximately three times in the early 2000s.

The causes of this collapse are structural. The United States has substantially underbuilt housing relative to population growth and household formation for fifteen years, a shortage estimated at three to four million units by most analyses. Zoning restrictions, building code costs, materials inflation, and labor shortages have combined to make new construction expensive and slow. Interest rate increases in 2022 and 2023 simultaneously made existing homeowners reluctant to sell

— because selling meant giving up historically low mortgage rates — and made new mortgages unaffordable for many potential buyers.

The result is a market where young families — the demographic that has historically been the primary source of first-time homebuyers — face barriers to entry that previous generations did not. The median age of a first-time homebuyer in the United States reached 38 in 2024, the highest on record. In the 1980s it was 29. The nine-year delay in first home purchase represents not just a personal financial setback but a nine-year delay in wealth accumulation, equity building, and the establishment of the financial foundation on which the rest of a family's economic life is typically built.

Institutional Investors: The Real Picture

Amid the housing affordability crisis, a specific narrative emerged with viral force: that Wall Street investment firms were buying up the country's single-family homes, pricing out ordinary buyers and converting ownership into permanent rentership. The story had real emotional power, and it pointed toward real phenomena — but it also contained significant inaccuracies that deserve careful examination, because the truth is both more complicated and, in important ways, more alarming than the simplest version of the story.

The most important factual correction: BlackRock, the world's largest asset manager, does not buy individual single-family homes. This claim circulated widely online and was repeated by politicians, but it is factually false. BlackRock invests in mortgage-backed securities and multifamily housing financing, but it is not the entity buying houses on your street.

Blackstone — an entirely separate company that shares a similar name — is a different matter. Blackstone is a private equity giant that founded Invitation Homes in 2012, following the 2008 financial crisis, specifically to acquire foreclosed single-family homes at scale and rent them out. It sold its Invitation Homes stake in 2019 for approximately

$7 billion. In 2024, Blackstone acquired Tricon Residential, adding nearly 38,000 homes to its portfolio and making it the third-largest institutional owner of single-family rentals in the country, with roughly 62,000 homes.

To be precise about the numbers: Blackstone owns approximately 0.06% of single-family homes in the United States. All institutional investors combined — entities that own 1,000 or more homes — own roughly 0.5% of the national single-family housing stock. These numbers do not support the claim that Wall Street has bought up a majority of the housing market.

But numbers at the national level can obscure what is happening at the local level — and that is where the real story is. In specific Sun Belt markets — Phoenix, Atlanta, Charlotte, Jacksonville, Tampa — investors have bought more than a third of available homes in certain price ranges during peak buying periods. In Atlanta's historically Black neighborhoods, researchers estimated that institutional investor activity cost potential homeowners approximately $4 billion in equity wealth between 2007 and 2016 by displacing individual buyers who would have built equity, with rental tenants who do not.

And there is a trajectory issue that deserves serious attention. If institutional ownership of single-family rentals grows from its current level of roughly 5% in certain markets to the projected 40% by 2030 — a projection cited in congressional testimony — the consequences for housing affordability, community stability, and generational wealth transmission would be profound. The question is not whether Wall Street already owns most housing. It is whether we want to wait until it does before deciding that this is a problem worth addressing.

> **DOCUMENTED:** In Sunbelt markets like Phoenix and Atlanta, institutional investors purchased more than one-third of available homes in certain price ranges during peak periods, paying cash and consistently outbidding

first-time homebuyers who require mortgage financing. Researchers found this activity cost Atlanta's homebuyers approximately $4 billion in lost equity wealth over a decade, with the impact concentrated in Black neighborhoods.

The RealPage Scandal: When Algorithms Fix Rents

Beyond institutional ownership, a separate but related scandal has exposed how the rental housing market has been manipulated against ordinary Americans through algorithmic coordination.

In August 2024, the U.S. Department of Justice filed an antitrust lawsuit against RealPage, a property management software company, alleging that RealPage's revenue management software had been used by major landlords to coordinate rent increases in a way that functioned as illegal price-fixing. The DOJ's complaint alleged that landlords using RealPage's algorithmic pricing tool were effectively sharing pricing information with competitors and deferring to the algorithm's rent recommendations, rather than competing independently on price.

The practical consequence, according to the lawsuit and multiple state attorney general investigations: rents in markets where RealPage was widely used rose significantly faster than they would have in a genuinely competitive market. Arizona's attorney general explicitly stated that RealPage and its co-defendants had contributed directly to Arizona's affordable housing crisis, with rents in Phoenix and Tucson rising at least 30% in large part because of what the lawsuit characterized as a rental monopoly.

The RealPage case illustrates something important: you do not need a conspiracy of billionaires secretly meeting in Davos to coordinate the extraction of wealth from ordinary people. You need an algorithm, widely adopted by dominant market players, that optimizes for their collective financial interests at the expense of tenants. The conspiracy, to the extent that word applies, is encoded in software and operates automatically, invisibly, at scale.

The Inheritance Problem: When Banks Get What Families Should

Perhaps the most profound and least discussed consequence of the housing affordability crisis is what it means for intergenerational wealth transfer. For most of American history, the home was the primary vehicle through which ordinary families transmitted wealth from one generation to the next. When parents died, the family home — paid off or substantially paid down after decades of mortgage payments — passed to their children. That asset, often representing hundreds of thousands of dollars, provided the next generation with a financial foundation that could be used to purchase their own home, start a business, fund education, or weather financial emergencies.

That mechanism is breaking down in two ways simultaneously. First, many people who would have been homeowners in previous generations are now renters, meaning they are building no equity and will have no home asset to transmit. Their monthly housing payments are building wealth — just not for them. They are building it for their landlords.

Second, for those who do own homes, the growing tendency to carry mortgage debt into late life — because homes were purchased later, refinanced more often, or used as ATMs through home equity loans during financial stress — means that homes are increasingly encumbered by debt when they pass to heirs. A home that is 60% mortgaged is a far less valuable inheritance than a fully paid-off home, and the interest paid over decades of mortgage debt represents an enormous wealth transfer from ordinary homeowners to financial institutions.

Add to this the growing prevalence of reverse mortgages — products that allow elderly homeowners to convert their home equity to cash while living in the home, with the bank claiming the property upon death — and you have a financial architecture that systematically redirects the accumulated equity of a lifetime from the family to the institution. The bank that financed the original purchase ends up with the house, and the children end up with nothing.

This is not the result of any single policy decision or corporate conspiracy. It is the cumulative outcome of decades of financial product development, policy choices, and structural changes in the housing market that, taken together, have restructured the flow of wealth in ways that consistently favor financial institutions over families and the present over the future.

The Subscription Economy and the Death of Ownership

The housing crisis is the most dramatic manifestation of the shift from ownership to access in the contemporary economy, but it is far from the only one. Across virtually every category of goods and services, the economic model of the past two decades has moved systematically away from purchase and toward subscription — from ownership toward rental, from possession toward access.

This shift has been presented, consistently and enthusiastically, by the companies driving it as a benefit to consumers. Lower upfront costs. No maintenance hassles. Always the latest version. Access to a vast library of content instead of a limited collection. The sharing economy. The circular economy. The service economy.

The marketing is compelling. The underlying economics tell a different story.

Software: You No Longer Own Your Tools

Adobe's Creative Suite — the industry-standard collection of software tools for graphic designers, photographers, filmmakers, and creative professionals — was once sold as a product. You bought it once. You owned it. You could use it as long as you wanted, on whatever computer you chose, with no ongoing financial obligation. A designer who purchased Photoshop CS6 in 2012 could still be using that same software in 2020 without paying another cent.

In 2013, Adobe discontinued the perpetual license model and moved entirely to a subscription service: Adobe Creative Cloud. The monthly fee that replaced the one-time purchase has made Adobe permanently, continuously dependent on subscription revenue from its users, and has made its users permanently, continuously financially dependent on Adobe. A designer who stops paying the monthly fee loses access to their tools immediately. The files they created with those tools may become inaccessible, because Adobe's proprietary formats require Adobe's software to open.

Adobe is not an exception. It is a template. Microsoft moved its Office suite to a subscription model. The video game industry has moved aggressively toward games-as-a-service models in which content that was once purchased is now rented through subscription services and can be removed at any time without the player's consent. The software tools that individuals and businesses depend on for their professional and personal lives are, increasingly, not owned. They are accessed — for as long as the company continues to provide the service and the customer continues to pay.

The difference between owning a tool and renting access to a tool is not merely semantic. It is a fundamental difference in the nature of your relationship to the things you depend on. When you own something, your continued use of it does not depend on an ongoing commercial relationship with the company that made it. When you rent access to something, your continued use depends entirely on that relationship — on the company's continued operation, continued decision to provide the service, continued willingness to serve you at a price you can afford, and continued compliance with whatever terms govern your use.

That dependency is not neutral. It is power. The company that controls access to the tools you need for your professional life has a form of leverage over you that the company that sold you a perpetual license does not have.

Media: From Library to Lease

The music collection that a person spent decades assembling — the albums purchased one by one, the vinyl records and CDs that represented both a financial investment and a cultural identity — has been replaced, for most people under forty, with a Spotify subscription. The movie collection is now Netflix, or HBO Max, or Disney+, or some rotating combination of platforms that contain the content you want to watch at any given moment and will contain different content next month.

The convenience is real. The library of available content is vastly larger than any individual could assemble through purchase. The cost per use, amortized over a subscription, is lower than purchasing each item individually. These are genuine advantages.

But the ownership advantages are gone. The music you listened to on Spotify this morning can disappear this afternoon if the licensing deal between Spotify and the rights holder expires. The movie you planned to watch this weekend can be removed from your streaming platform without notice. The e-book you 'purchased' from Amazon is not owned — it is licensed, and Amazon has the technical and legal ability to remove it from your device without refund, which the company famously did in 2009 when it remotely deleted copies of George Orwell's 1984 and Animal Farm from customers' Kindle devices due to a licensing dispute.

The specific, bitter irony of a corporation remotely deleting 1984 from people's personal devices should not be lost. But beyond the irony, the incident illustrated, in practical and visible terms, the fundamental difference between owning a book and licensing access to a book's content through a digital intermediary that can revoke that access at its discretion.

When the things you depend on — your entertainment, your tools, your professional software, your books, your music — are accessed rather than owned, the control you have over your own cultural and intellectual life diminishes in ways that compound invisibly over time.

The Subscription Debt Trap

The aggregate financial cost of the subscription economy has received surprisingly little attention relative to its impact on household finances. The average American adult now holds subscriptions to multiple streaming services, a cloud storage service, a music service, a software suite, a gym membership, a news subscription, and dozens of smaller recurring charges for apps and services that accumulate on credit cards and bank statements in amounts too small to notice individually but significant in aggregate.

Research by financial services companies has consistently found that people dramatically underestimate their monthly subscription spending. A 2022 survey by C+R Research found the average American spends approximately $219 per month on subscriptions — far more than most respondents estimated. More strikingly, 42% of respondents said they were still paying for subscriptions they had forgotten about entirely.

The subscription model is specifically designed to exploit the psychological quirks that make recurring small charges easy to ignore and difficult to cancel. Free trials that convert to paid subscriptions. Annual billing that creates a large, infrequent charge that does not register in monthly budget calculations. Cancellation processes deliberately made difficult and time-consuming to reduce churn. The economic engineering of the subscription model is, from the consumer's perspective, a machine for generating invisible recurring expenditure.

For younger generations carrying student loan debt, facing high housing costs, and navigating stagnant wages in many sectors, the subscription economy represents a quiet but significant drain on the financial resources that might otherwise be accumulating toward the assets — the home, the savings, the investments — that previous generations used to build generational wealth.

Central Bank Digital Currencies: The Most Consequential Financial Debate You've Never Had

Of all the financial developments discussed in this chapter, the one with the most profound long-term implications for ordinary people — and the one that has received the least proportionate public attention — is the global movement toward central bank digital currencies, or CBDCs.

As of 2025, according to the Atlantic Council's CBDC Tracker, 137 countries and currency unions representing 98% of global GDP are exploring a CBDC. The Bahamas, Jamaica, and Nigeria have launched them. China's digital yuan pilot has reached 7 trillion yuan in total transactions across 17 provincial regions. The European Central Bank is in active preparation for a digital euro. The Bank of England is developing a digital pound. India's digital rupee pilot has grown 334% in a single year.

The United States, notably, has moved in the opposite direction. President Trump issued an executive order in January 2025 prohibiting federal agencies from taking any action to establish, issue, or promote a retail CBDC. The House passed the Anti-CBDC Surveillance State Act. Bipartisan concern about the surveillance and control implications of a programmable digital dollar has, at least for now, halted American CBDC development.

But the United States represents one country. The global financial architecture is shifting. Cross-border CBDC systems — including China's mBridge project linking multiple countries' digital currencies — are developing infrastructures that will increasingly define how international commerce and finance operate. The question of whether a digital dollar eventually exists, in some form, is not closed by the current political environment. And the question of what properties it would have — who can see your transactions, who can restrict your spending, what happens to financial privacy in a fully digital currency world — is one of the most consequential policy questions of the next decade.

What Makes a CBDC Different from the Digital Money You Already Use

A common reaction to CBDC discussions is: I already use digital money. I use my debit card, I pay with my phone, I transfer money through apps. How is a CBDC different?

The difference is architectural, and it is fundamental. When you use your debit card, your transaction passes through a commercial bank — an intermediary that holds your deposit, processes the transaction, and maintains a relationship with you. The government does not have direct visibility into individual transactions, though it has access through subpoenas and regulatory oversight. The commercial bank layer provides a degree of privacy buffer between your spending and direct government surveillance.

A CBDC eliminates that buffer. In a CBDC system, digital currency is a direct liability of the central bank — not of a commercial bank. The central bank, or a government-designated authority, would have direct, real-time visibility into every transaction made with CBDC. Every purchase you make, every payment you send, every financial action you take with the currency becomes a data point in a government-accessible ledger.

The Wikipedia entry on CBDCs, drawing from a range of expert analyses, captures the surveillance concern directly: CBDCs would give governments 'direct visibility of financial transactions' — an eagle-eyed view on the spending of everyone. Digital currency would give a country 'broad new powers when it comes to surveillance and controlling its population.' Data from tracing money routes could lead to losing financial privacy.

But surveillance is only one dimension of the concern. The other is programmability.

Programmable Money: The Power to Condition Your Spending

Unlike physical cash — which can be spent on anything by anyone without any system recording it or restricting it — digital currency can be programmed. A central bank or government can build rules into the currency itself that govern how it may and may not be used.

This is not theoretical. China's digital yuan has been tested with expiration dates — money that must be spent by a certain date or it disappears, forcing spending and preventing saving. The technical architecture of CBDCs allows for restrictions on what categories of goods can be purchased, restrictions on transactions above certain amounts, restrictions based on the identity or status of the parties involved, and restrictions that can be imposed or lifted instantaneously by whoever controls the programming.

Senator Ted Cruz, in introducing the CBDC Anti-Surveillance State Act, described the concern plainly: a CBDC is government-controlled programmable money that could give the federal government not only significant transaction-level data but also the ability to program the CBDC to choke off politically unpopular activity.

The progressive critique of the same technology runs parallel: a CBDC could be programmed to cut off political donations to disfavored organizations, to restrict purchases of firearms or certain medications, to freeze the accounts of activists or protesters, to implement social benefit conditionality — you can only spend your government assistance on approved categories of goods. The political valence of the concern differs, but the underlying technical concern is identical: programmable money is power, and whoever programs the money has power over the people who use it.

China's social credit system, which we examined in Chapter Two, demonstrates what this convergence of financial surveillance and behavioral consequence can produce. China's system uses digital payment infrastructure, among other data sources, to generate scores that determine individuals' access to services and opportunities. The digital yuan extends this surveillance infrastructure to currency itself — making spending behavior not just observable but directly manageable by the state.

Western governments are not China. Democratic institutions provide protection against the most egregious abuses of programmable currency. But the protections of democratic institutions are not permanent. They depend on political will, on the health of democratic norms, on the vigilance of citizens and journalists and courts. The infrastructure that would enable financial surveillance and control, once built, does not disappear when political conditions change. It persists, waiting for conditions under which it will be used.

> **PROGRAMMABLE CURRENCY RISK:** CBDC expert analyses cited in Wikipedia's CBDC entry warn that digital currencies could 'prohibit the purchase of alcohol on a weekday,' 'make it impossible to donate to a vocal NGO,' and enable governments to fine citizens 'in a split second for behaviors deemed undesirable.' These are not worst-case fantasies — they describe technical capabilities that exist in current CBDC designs.

The Cashless Society: Privacy Lost in Small Steps

The CBDC debate is partly academic in the United States in the current political environment. But the broader movement toward a cashless society — driven by the convenience of digital payments and the interests of financial companies — is proceeding regardless of

CBDC policy, and it raises many of the same privacy and control concerns in a less centralized but equally consequential form.

Physical cash is, at the moment, the only form of money that provides complete transaction privacy. When you pay for something in cash, the transaction is not recorded in any system accessible to your bank, your government, or data brokers. Cash is the financial equivalent of a private conversation — the only truly private conversation in an increasingly surveilled economy.

The move away from cash is accelerating rapidly. Many businesses in major American cities have gone cashless, citing the cost and friction of cash handling. The COVID pandemic dramatically accelerated digital payment adoption, with many merchants refusing cash for hygiene reasons that long outlasted their practical justification. Younger generations use physical currency less and less, not through any deliberate choice but through the quiet normalization of card and phone payments as the default mode of transaction.

Each step away from cash is a step toward a financial record of your life. Your debit card history knows every restaurant you have visited, every pharmacy you have used, every store where you have shopped, every bar where you have had a drink. Your credit card history is a diary of your consumption. Your bank account records are a map of your financial life. None of this information is perfectly private even today — banks comply with law enforcement requests, data brokers buy transaction data, and the surveillance economy we described in Chapter One extends into financial behavior as thoroughly as it extends into any other domain.

But cash provides an exit. It provides a space in which a transaction can occur without a record, without surveillance, without the transaction being added to a profile that follows you through a digital economy. The elimination of cash — whether through explicit policy, commercial pressure, or the simple inconvenience of cash in an increasingly digital world — would eliminate that exit permanently.

The European Digital Euro: A Preview of the Debate

The European Central Bank's digital euro project offers the clearest window into what a CBDC designed by a democratic government might look like — and what the debates that surround its design reveal about the genuine tensions between the legitimate goals of digital currency and the privacy concerns its critics raise.

European policymakers have been explicit that privacy is a design requirement for the digital euro. The ECB has proposed a system in which low-value, in-person transactions would be conducted without real-time central bank visibility — a deliberate attempt to preserve something like the privacy of cash for everyday purchases. The ECB's design contemplates holding limits — a cap on how much digital euro any individual can hold — specifically to prevent the digital euro from displacing commercial bank deposits in a way that would create systemic financial risk.

These are genuine attempts to address legitimate concerns, and they reflect the fact that democratic governments designing CBDCs face accountability pressures that authoritarian governments do not. But the history of privacy protection in digital systems should inspire caution rather than confidence. The data minimization principles written into GDPR have been honored inconsistently. The privacy protections built into smartphone operating systems have been eroded by commercial and governmental pressure over time. The promise that a system will be designed with privacy in mind does not guarantee that the system will remain private as political conditions, commercial pressures, and technological capabilities evolve.

The Pension Betrayal and the Retirement Savings Illusion

The third pillar of ordinary wealth — the guaranteed retirement income that defined-benefit pensions provided — has been almost entirely dismantled for private sector workers over the past forty years. This transformation is so complete and has been so thoroughly normalized that younger workers may not fully understand what was lost, or that what currently exists is a significantly inferior substitute.

From Guaranteed to Gambled

In 1979, approximately 62% of private sector workers with employer-sponsored retirement plans had defined-benefit pension plans — plans in which the employer contributed to a pension fund and guaranteed the employee a specific monthly payment in retirement for the rest of their life. The employer bore the investment risk. If the pension fund's investments underperformed, the employer's obligation did not diminish. The employee's retirement income was guaranteed regardless of what happened in financial markets.

By 2019, fewer than 4% of private sector workers were covered by defined-benefit pensions. The near-universal replacement is the 401(k) and similar defined-contribution plans, in which employees contribute their own money (sometimes matched by employers) to individual accounts that are invested in financial markets. The employee bears the investment risk entirely. If markets perform well, retirement savings grow. If markets perform poorly — as they did in 2001-2002, 2008-2009, and with significant volatility since — retirement savings shrink. If a worker retires in a market downturn, they may lose a substantial portion of their accumulated savings at the worst possible moment.

This shift from defined-benefit to defined-contribution plans was the largest involuntary transfer of financial risk in modern American history, and it was accomplished so gradually, over so many years, with so little explicit public debate about its implications, that it passed largely without the outrage it deserved. Corporations shed pension obligations that represented significant balance sheet liabilities. The risk that those obligations represented — the risk that markets would underperform, that workers would live longer than expected, that the fund's assets would prove insufficient — was transferred entirely to workers, who were generally not told clearly what they were taking on.

The 401(k) Was Never Designed for This

The 401(k) provision of the Internal Revenue Code, from which the most common American retirement savings vehicle takes its name, was not originally designed as the primary retirement savings mechanism for the American workforce. It was designed as a supplemental savings tool — a way for executives and highly compensated employees to defer additional income beyond their pension. Its author, Pennsylvania benefits consultant Ted Benna, has himself expressed regret at how the provision was repurposed to replace rather than supplement defined-benefit pensions.

The consequences of trying to make a supplemental savings vehicle serve as a primary retirement system are visible in the data. As of 2023, approximately 57% of American adults have no retirement savings at all — not a 401(k), not an IRA, nothing. Of those who do have retirement savings, the median balance is dramatically insufficient to fund a comfortable retirement. The National Institute on Retirement Security has found that two in three working Americans are not on track for a financially secure retirement.

The retirement savings crisis falls most heavily on those with the least capacity to bear it. Workers in low-wage jobs rarely have access to employer-sponsored retirement plans. Workers who change jobs frequently — as has become the norm in the gig economy and project-based work economy — often cash out small 401(k) balances rather than rolling them over, eliminating years of compounding. Workers who face financial emergencies — medical expenses, job loss, housing instability — often raid retirement savings as a last resort, paying penalties and taxes that further erode the balance.

The people most likely to arrive at retirement with inadequate savings are also the people most likely to have depended on Social Security as their primary retirement income, to have built home equity as their primary asset, and to have expected to leave something to their children. The convergence of inadequate retirement savings, declining home equity transmission, and the subscription-debt economy is not hitting Americans evenly. It is concentrating most heavily on those who can least afford it.

The Death Tax Debate and the Estate That Never Was

In historical context, inheritance was one of the primary mechanisms through which ordinary families transmitted wealth across generations. Land passed from parents to children. The family business was handed down. The paid-off home was inherited debt-free. The gold coins buried under the floorboards — literal, in the case of many immigrant families who did not trust banks — were found and used by the next generation.

The modern financial system has inserted itself into this transmission mechanism at multiple points, capturing wealth that previous generations would have passed to their heirs. Reverse mortgages capture home equity. Estate planning costs extract fees from the transmission process. Probate proceedings, where assets are not properly structured, can consume significant portions of modest estates. The federal estate tax, though now applicable only to estates above $13.6 million, interacts with state estate taxes in ways that affect more families. And the debt that modern families carry — student loans, mortgage debt, medical debt, credit card debt — often means that the estate being transmitted is a net negative rather than a positive transfer.

The single most significant development undermining inheritance for ordinary families is not the estate tax, which affects relatively few. It is the shift from home equity as the primary asset of working-class estates to home equity encumbered by ongoing debt, or absent entirely because the family never achieved homeownership. When the home is not owned, the primary mechanism for intergenerational wealth transfer in working-class families does not exist. The rent payments that accumulated over a lifetime created no asset. The landlord benefited. The children inherited nothing.

The Wealth Concentration Machine: Who Benefits When You Don't Own

Every shift from ownership to access, from defined benefit to defined contribution, from cash to digital, from homeownership to rentership has a mirror image: it transfers wealth from the many to the few. Understanding who benefits from each of these transitions is essential to understanding why they have proceeded so relentlessly and with so little effective resistance.

When You Rent, Someone Else Builds Equity

Every rent payment a tenant makes is a mortgage payment made by someone else — the landlord. The tenant pays for housing but builds no equity. The landlord provides housing and builds equity with every tenant's payment. Over a lifetime of rentership, a person may spend a sum equivalent to the cost of a home without ever owning one. The total spent on rent is indistinguishable in amount from the total spent on mortgage principal and interest, but the outcomes are radically different. The homeowner ends their life with an asset. The renter ends their life with receipts.

The institutional investors who have entered the single-family rental market at scale understand this arithmetic with great clarity. Their business model is precisely the conversion of what would have been owner-occupied housing — housing in which individual families would have been building equity — into housing in which they receive equity-building rent payments from families who will never own. The fact that this does not require any malicious intent is not a comfort. The structural consequences are the same regardless of intent.

The Financial Industry's Interest in Your Retirement Uncertainty

The shift from defined-benefit pensions to defined-contribution plans created an enormous new market for the financial services industry. Assets under management in defined-contribution plans in the United States now exceed $10 trillion. The fee income generated by managing those assets — even at the seemingly modest rates that characterize index funds and most 401(k) options — amounts to tens of billions of dollars annually.

The financial industry has a direct, structural interest in the continuation of a retirement system in which individuals manage their own savings through financial products. This interest is expressed through lobbying against policies that would strengthen Social Security, against policies that would create public pension alternatives to private 401(k) plans, and against fee regulations that would restrict the income generated from managing retirement assets.

This is not conspiracy. It is rational self-interest, expressed through the normal channels of political influence in a democracy where campaign contributions and lobbying are legal. But it means that the people with the most at stake in the continuation of the current system — which transfers both wealth and risk from ordinary workers to financial institutions — are also the people with the greatest capacity to influence the political decisions that would change it.

Big Tech's Interest in the Cashless Society

Apple, Google, PayPal, Stripe, Visa, Mastercard — the companies that provide the infrastructure for digital payments have a direct financial interest in the decline of cash. Every digital transaction generates a fee that is absent from cash transactions. The payment processing companies earn interchange fees on card transactions that average 1.5-3.5% of the transaction value. On the $10 trillion in consumer spending that flows through digital payment systems annually, those percentages represent hundreds of billions of dollars in fee income.

The data generated by those transactions is equally valuable. We examined in Chapter One the extent to which transaction data forms part of the behavioral profiles that fuel the surveillance economy. Payment processors, banks, and the data brokers who purchase transaction data all have commercial interests in knowing what you buy, where, when, and how much you spend. Cash generates none of this data. Every dollar that moves from cash to digital payment is a dollar that generates both fee income and behavioral data for the financial technology industry.

The commercial interests of the payments industry are not the reason cash is declining. Convenience is the primary driver of digital payment adoption, and the convenience is genuine. But the structural alignment between the commercial interests of technology companies and the policy direction toward a cashless society is worth noting, because it means that the path of least political resistance — doing nothing while the market moves toward cashless payments — happens to benefit some of the most politically influential companies in the world.

The Debt Architecture of Modern Life

Underlying all of the wealth dynamics discussed in this chapter is a debt architecture that has become so normalized it is nearly invisible. The contemporary American economy is structured around the assumption that ordinary people will carry significant debt for most of their productive lives — student debt, mortgage debt, auto debt, credit card debt — and that the management of that debt, through payment and refinancing and consolidation and the occasional bankruptcy, will constitute a significant portion of their financial activity.

This is a historically unusual state of affairs. For most of human history, the accumulation of significant personal debt was associated with crisis — war, famine, catastrophic loss. Debt was something you acquired under duress and escaped as quickly as possible. The concept of debt as a normal, manageable, permanent feature of middle-class financial life is a relatively recent invention, aggressively marketed by the financial industry that profits from it.

The Student Loan Architecture

The student loan system in the United States represents one of the most consequential financial policy decisions of the past fifty years — the decision to fund the expansion of higher education access through individual debt rather than collective public investment. The consequences of that decision are now visible in a generation of Americans who entered adulthood carrying balances that have, in many cases, grown larger rather than smaller despite years of payments, due to interest capitalization and income-based repayment structures that extend debt across decades.

Total outstanding student loan debt in the United States exceeded $1.77 trillion in 2024 — more than all credit card debt and auto loan debt combined. The average graduate of a four-year program carries approximately $37,000 in student debt at graduation, a figure that climbs significantly for graduate and professional degree holders. For the many borrowers who attended for-profit institutions or did not complete their degrees, debt without the income premium of a credential is a particularly severe burden.

The student debt burden directly suppresses the wealth-building activities that previous generations took for granted. Research consistently shows that high student debt is associated with delayed home purchase — the primary wealth-building activity of working and middle-class families — delayed family formation, reduced retirement savings contributions, and reduced entrepreneurial activity. The debt that was supposed to be an investment in human capital has, for millions of Americans, functioned instead as a mortgage on their future.

The financial institutions that hold student debt — including the federal government, which now holds the majority of outstanding student loans, as well as private lenders — do not face the consequences of this burden. Their loans are either guaranteed or held by institutions with the scale to absorb defaults. The risk of the student debt system is borne entirely by individual borrowers, many of whom took on that debt at eighteen without the life experience or financial sophistication to understand what a $50,000 or $100,000 debt obligation would mean for their financial life in their twenties and thirties.

Medical Debt: The Uniquely American Catastrophe

No other wealthy nation generates medical debt the way the United States does. The combination of a predominantly private insurance system, high out-of-pocket costs, and a billing complexity that is genuinely incomprehensible to most patients creates a system in which a serious medical event — a hospitalization, a cancer diagnosis, a traumatic accident — can generate debt that rivals a mortgage in scale, appearing with no warning and offering no opportunity for advance planning.

Approximately 100 million Americans carry medical debt, according to a 2022 investigation by KFF Health News and NPR. Medical bills are the leading cause of personal bankruptcy in the United States. The debt does not discriminate by income — a middle-class family with insurance can be devastated by a single serious illness that exceeds out-of-pocket maximums, results in balance billing from out-of-network providers, or generates costs in the months before insurance coverage begins.

Medical debt is a uniquely efficient destroyer of savings and assets. Because medical crises are unpredictable and can generate large costs rapidly, they are impossible to plan for with ordinary savings. A family that has spent years building home equity and retirement savings can find that financial foundation eroded or eliminated by a single serious illness. The wealth that was supposed to flow to the next generation flows instead to medical providers and the debt collectors who purchase medical debt at pennies on the dollar and pursue it aggressively.

Credit Card Debt: The Invisible Tax on Poverty

Credit card debt carries interest rates that, in a regulatory environment with any meaningful consumer protection, would be prosecuted as usury. The average credit card interest rate in the United States in 2024 exceeded 21% annually. For the tens of millions of Americans who carry revolving balances — who cannot pay off their full balance each month because their income does not cover their expenses — this rate represents a significant and continuous transfer of wealth from households to financial institutions.

A family carrying $10,000 in credit card debt at 21% and making only minimum payments will pay approximately $12,000 in interest over the life of the debt before the principal is retired. They will pay more than twice what they borrowed. The money that flows to the credit card company in interest payments is money that does not go toward home down payments, retirement savings, or children's education. It is wealth extraction in its most direct form — money removed from family balance sheets and added to institutional ones.

The credit card industry is not a neutral service provider. It is a designed system, with extensive research behind its design, to maximize the duration and size of revolving balances. Low minimum payments that ensure debts are repaid over years rather than months. Reward programs that encourage spending above what cash-flow would support. Zero-percent introductory offers that trap users into high-rate balances when the promotional period expires. These are not accidental features. They are profit mechanisms, and they function exactly as designed.

What 'You Will Own Nothing' Actually Means — Separating Fact from Fear

We began this chapter with eight words from a WEF video essay that became a cultural flashpoint. It is worth returning to those words now, with the full context of the chapter behind us, to ask: what does the observable trajectory of financial trends actually suggest about the ownership future of ordinary Americans?

The honest answer is not that a secret global elite is implementing a plan to strip people of property. It is more complicated and, in some ways, more alarming than a conspiracy, because it does not require conspiracy to proceed.

The forces driving ordinary people toward less ownership are not coordinated by a central authority. They are the cumulative output of thousands of individual decisions by corporations, financial institutions, landlords, technology companies, and policymakers, each acting in their own rational self-interest in ways that collectively create a system that concentrates ownership upward. No one decided to make housing unaffordable for young families. It is the emergent outcome of zoning laws, supply constraints, institutional investment, and financial product design, each element of which was created by actors with coherent, often defensible rationales.

No one decided to make retirement insecure for working Americans. It is the outcome of corporate decisions to shed pension obligations, congressional inaction on Social Security reform, financial industry lobbying against public retirement alternatives, and individual workers' inability to save at the rate required to fund a comfortable retirement on their own.

No one decided to make inheritance impossible for most families. It is the outcome of housing unaffordability, debt accumulation, the financial products that capture home equity in late life, and the structural barriers to wealth accumulation that make many families net debtors rather than net asset holders at the end of life.

The result — a world in which ordinary people own less, control less, and pass less to the next generation — does not require eight powerful men meeting in Davos to design. It has been designed, piece by piece, by the ordinary operation of a political economy in which the interests of financial institutions have consistently prevailed over the interests of ordinary people, not through dramatic villainy but through the quiet, relentless application of economic and political power.

The Generational Stakes: What Is Actually Being Lost

The deepest concern about the trajectory described in this chapter is not personal financial hardship, though that hardship is real. It is the erosion of the material foundation for the kind of independence, security, and intergenerational opportunity that free societies have historically associated with full participation in economic life.

A person who owns their home has a form of security that cannot be revoked by a landlord's decision to sell or raise rents. A person who controls their financial assets without constant intermediary fees and surveillance has a form of privacy and autonomy that is not dependent on the goodwill of financial institutions. A person who passes meaningful assets to their children provides those children with options — educational, entrepreneurial, residential — that children who inherit nothing do not have.

These are not luxuries. They are the material conditions for the kind of life that democratic societies claim to value: security, independence, opportunity, the ability to plan for a future and pass its fruits to those who come after. When those conditions erode for the majority while concentrating in the hands of a minority, the democratic promise is not kept.

The conversation about CBDCs, about institutional home ownership, about the subscription economy and the retirement savings gap, is ultimately a conversation about what kind of society we want to live in and whether we are willing to make the political choices necessary to build it. Those choices require citizens who understand what is happening to the material conditions of their lives — not just in the dramatic terms of state surveillance and cyber warfare, but in the quieter, daily terms of who owns what, who controls what, and who benefits when ordinary people have less.

What You Can Do: Financial Autonomy in a Hostile Landscape

The structural forces described in this chapter are real and powerful. They cannot be countered by individual action alone. But individual action, informed by understanding of the threat, is the necessary foundation for any broader response.

Prioritize ownership over access wherever economically feasible. Where you have the choice between purchasing and subscribing — for software, for media, for tools — understand the long-term cost difference and the control difference. Ownership, even at higher upfront cost, provides a form of security that access does not. Not every ownership decision is financially optimal in the short term. Over a lifetime, the accumulation of owned assets versus rented access has profound consequences.

Understand your credit as a strategic resource. Credit card debt at 21% interest is not neutral financing. It is an ongoing transfer of wealth from your household to a financial institution. If you carry revolving balances, the single most high-leverage financial decision available to most families is eliminating that debt as aggressively as possible. The interest saved on $10,000 in credit card debt paid off over one year rather than ten is more than the typical annual return on retirement investments.

Take retirement savings seriously early. The mathematics of compound interest are unambiguous: money saved in your twenties is worth dramatically more in retirement than money saved in your forties. Every year of delay in retirement saving has compounding consequences. The 401(k) system is imperfect, the risks are real, and the guarantees that defined-benefit pensions provided are gone — but contributing to whatever retirement savings vehicle is available, at the maximum affordable level, from as early as possible, remains the most reliable path to retirement security for most Americans.

Stay informed about CBDC and digital currency policy. The decisions being made now about the architecture of digital financial infrastructure will define the financial privacy landscape for decades. Participate in the public debates. Understand what your representatives have said and done about CBDC legislation. Understand the distinction between a privacy-respecting digital payment system and a surveillance-capable programmable currency. These are not obscure technical questions. They are questions about how much control you will have over your own financial life.

Demand housing policy that serves families, not investors. The homeownership gap, the rental housing affordability crisis, and the concentration of single-family housing in institutional hands are all addressable through policy. Zoning reform that enables more housing construction, restrictions on institutional purchases of single-family homes, support for first-time homebuyer programs, and policies that strengthen tenants' rights in the face of institutional landlords are all available options. They require political will that is generated by citizens who understand the stakes and vote accordingly.

* * *

END OF CHAPTER FOUR

Continue to Chapter Five: The Children We Are Raising

CHAPTER FIVE

The Children We Are Raising:

What We Are Handing Them, What We Are Taking From Them, and the World We Could Still Build

---

"Every generation imagines itself to be more intelligent than the one that went before it, and wiser than the one that comes after it." George Orwell

The Baby Who Was Already a Data Point

Before she had a name officially registered on a birth certificate, she had a digital footprint.

Her mother's pregnancy announcement posted at twelve weeks on Instagram, 847 friends plus public included her first ultrasound image. Her name revealed at twenty weeks in a gender reveal video that generated forty-three comments and two hundred and twelve likes. The hospital photo at birth, tagged with the city and the hospital name. The first week of daily updates: what she weighed, how she slept, whether she was eating well, which side of the family she resembled. All of it public. All of it permanent. All of it harvested.

By her first birthday, this child who had never operated a device, never agreed to a terms of service, never made a single conscious decision about her digital presence had a data profile that included her full name, date of birth, physical location, the names and faces of her parents and siblings, the hospital where she was born, her weight and feeding patterns, her temperament and developmental milestones, the financial status of her family inferred from the neighborhood and products visible in the photos, and hundreds of photographs of her face at every angle and developmental stage that any facial recognition system would ever need to identify her.

This is not an unusual story. This is the story of nearly every child born in the developed world since approximately 2010. It is, in the words of researchers, the creation of a child's first digital dossier an involuntary data record assembled by loving parents who were thinking about sharing joy with family and friends and were not thinking, because almost no one told them to think, about what that sharing would mean for the person in the photographs.

That person could not consent. Could not object. Could not choose. Could not imagine, from within the warm, safe blur of infancy, that the images being captured would follow them through job applications and college admissions and credit evaluations and political campaigns and romantic relationships that had not yet been imagined. Could not know that by the time they were old enough to understand what a digital footprint was, theirs would already be decades old and effectively impossible to erase.

We have handed an entire generation of children something that no generation in human history has ever received: a permanent, searchable, algorithmic record of their entire existence, assembled before they could walk, by people who loved them, for reasons that were almost entirely good, with consequences that are only now beginning to be understood.

This chapter is about those consequences. It is about what the digital world is doing to our children to their privacy, their mental health, their developing identities, their futures. It is the most personal chapter in this book, because the children we are talking about are not abstractions. They are your children. Your grandchildren. Your nieces and nephews and the kids on your street and in your school district and in your community. They are the future in the most literal sense the human beings who will inherit whatever world we are building, or failing to build, right now.

And then it is about something else. Something this book has been building toward from the beginning the possibility, real and documented and grounded in observable human determination, that we can choose differently. That the generation growing up in the age of surveillance and AI and deepfakes and data harvesting could instead grow up in an age of digital rights, genuine privacy, and technology that serves human flourishing rather than extracting from it.

That is not a fantasy. It is a choice. And every chapter in this book has been building the case that it is a choice we can still make.

Born Into the Data Economy: The Sharenting Crisis

The word 'sharenting' a portmanteau of sharing and parenting was coined to describe something that needed a name because it had never existed before: the practice of parents routinely publishing detailed information about their children's lives on social media platforms, from infancy onward, without the child's knowledge or consent.

The scale of the phenomenon is staggering. Research conducted in 2010 found that 92% of American two-year-olds already had an online presence. More recent data suggests the situation has only deepened since. Of those children with an online presence, one-third were first posted on the date of birth itself. Approximately half of those birth-day posts included the child's first name, their physical location through geo-tagging, and enough contextual information hospital name, neighborhood, family details to construct a basic identity profile.

On day one of life, before a child has drawn breath in the outside world for twenty-four hours, an unknown number of strangers, data brokers, and algorithmic systems have already begun to know them.

What Parents Share Without Realizing

Most parents who share information about their children online are not thinking about data brokers or facial recognition algorithms. They are thinking about grandparents who live three states away and want to see the baby. They are thinking about the community of other parents who understand the specific exhaustion and joy of new parenthood. They are thinking about preserving memories, maintaining social connection, sharing the extraordinary ordinary experience of watching a small person become themselves.

These are not bad motivations. They are thoroughly human motivations. And the technology platforms that profit from sharenting have been carefully designed to feel like natural, benign channels for exactly these impulses. The architecture of social media the frictionless share button, the hearts and comments that reward posting, the algorithmic amplification of family content makes sharing feel like the obvious, natural, socially rewarded thing to do. Not sharing, by contrast, feels effortful, antisocial, even a little paranoid.

But the information being shared adds up in ways that parents rarely calculate. A 2021 analysis found that the average parent shares approximately 1,000 images of their child online before that child's fifth birthday. Across those 1,000 images are embedded: the child's face at dozens of developmental stages, sufficient for facial recognition training and future identification. Geographic data from photo metadata or background details identifying home, school, and frequently visited locations. Names, ages, and faces of family members and friends. Health and developmental information from updates about milestones, illnesses, and doctor visits. Behavioral and temperament data from descriptions of the child's personality and emotional patterns. Financial and social status data inferred from the visible home, vehicles, clothing, and activities.

Barclays Bank, in a 2018 report, predicted that sharenting would account for two-thirds of identity fraud cases by 2030. The mechanism is straightforward: the name, date of birth, and home address shared in birth announcements and early posts represent exactly the data needed to establish a fraudulent identity. That data, stored in commercial databases and accessible to data brokers, will be available when that child reaches adulthood and begins to interact with financial systems that rely on the same identifying information.

A child born in 2015 will turn eighteen in 2033. The data that was shared in the first weeks of their life will be ten years old. But it will still exist, in the same databases, ready to be used against them by identity thieves who have been storing it, patiently, waiting for the moment it becomes financially exploitable.

BARCLAYS 2018 PREDICTION: Sharenting parents sharing children's information on social media is projected to account for two-thirds of identity fraud cases by 2030. The name, birth date, and location routinely shared in birth announcements contain exactly the data needed to construct a fraudulent identity when the child reaches adulthood.

The Right to a First Chapter

Beyond the practical risks of identity fraud and data exploitation, sharenting raises a deeper question that has received less attention but deserves more: does a child have the right to write their own story?

Every human being, in every culture and every era, has had the experience of constructing a self of deciding, through accumulation of choices and experiences, who they are and who they present themselves to be. The teenager who went through an awkward phase and grew out of it. The young adult who held embarrassing opinions and changed their mind. The person who made mistakes and learned from them without those mistakes being permanently searchable.

The ability to evolve to be different at twenty-five than you were at fifteen, to leave behind the worst moments of your developing years, to present yourself to the world as the person you have become rather than the composite of everything you have ever been is fundamental to the human experience of growing up. It is not just a psychological luxury. It is a social and professional necessity.

Sharenting strips that right from children before they are old enough to understand what is being taken. The child who threw a spectacular tantrum at a birthday party at age three, whose meltdown was filmed and shared and received hundreds of laughing emojis from well-meaning relatives, did not consent to that moment being their introduction to the internet. The teenager who finds that their embarrassing elementary school photos are the first results when their name is searched has inherited a digital history they did not write and cannot delete.

Legal scholars have begun to develop frameworks for what they call 'the right to be forgotten' for children the idea that the digital record created about a person during childhood, before they had the capacity to consent, should be erasable at their request when they reach adulthood. France and several other European countries have moved toward legal protection of this right. In the United States, no such comprehensive protection currently exists.

Illinois, recognizing a specific dimension of this problem, amended its child labor laws in 2024 to require that parents who feature their children in monetized social media content as influencers or in family YouTube channels set aside a portion of the earnings in a trust for the child, who cannot access the funds until they turn eighteen. The law protects children from having their labor and likeness commercially exploited by the very parents who are supposed to protect them. It is a meaningful step. But it addresses the monetization problem while leaving the broader sharenting problem largely untouched.

The Algorithm That Is Raising Your Children

If sharenting represents what parents are doing to their children's digital futures, the social media algorithm represents what the platforms are doing to children's developing minds and the picture painted by the accumulated research is one of the most alarming in the field of child development in decades.

The U.S. Surgeon General issued a formal advisory in 2023 an extraordinary document, the first of its kind from the nation's top public health official warning that social media poses a significant risk to the mental health of children and adolescents. It was not a cautious, hedged, both-sides document. It was an alarm.

The following year, in June 2024, Surgeon General Vivek Murthy went further, calling for warning labels on social media platforms comparable to the warnings on cigarettes and alcohol the first time in American history that a Surgeon General had called for such labels on a technology product. The comparison to tobacco was not rhetorical flourish. It was a deliberate invocation of a public health crisis that also took decades to name, during which children were the most vulnerable victims and corporate interests were the primary obstacle to protection.

The Numbers Behind the Warning

The Surgeon General's 2023 advisory drew on a substantial and converging body of research. Up to 95% of young people ages 13 to 17 report using a social media platform. More than a third say they use social media almost constantly not daily, but constantly, meaning they are checking, scrolling, responding throughout every waking hour. Nearly 40% of children between the ages of 8 and 12 are also using social media, despite minimum age requirements of thirteen on most platforms.

The health data correlated with this usage is stark. Children and adolescents who spend more than three hours per day on social media face double the risk of experiencing poor mental health outcomes, including symptoms of depression and anxiety. This is not a marginal effect. Doubling the risk of depression and anxiety in teenagers is a public health crisis by any reasonable measure. And teenagers in grades eight and ten were, as of the Surgeon General's advisory, spending an average of three and a half hours per day on social media already above the threshold where risk doubles.

Teen depression began rising around 2012. Smartphone penetration crossed 50% in the American population in approximately 2012. The likes button became ubiquitous on major platforms in the same period. Social media algorithms became substantially more sophisticated in their psychological targeting around the same period. The correlation is not proof of causation. But as Professor Jean Twenge of San Diego State University, one of the leading researchers in this field, stated: for all of these inflection points to coincide and for it not to be related to each other strains the bounds of coincidence beyond any reasonable interpretation.

The mental health effects are not distributed equally. Girls are substantially more affected than boys, particularly regarding body image, eating disorders, and depression related to social comparison. Instagram with its emphasis on appearance, lifestyle curation, and visual self-presentation was ranked in a Royal Society for Public Health study as the single most harmful social media platform for young people's mental health. TikTok's algorithmic engine, which is among the most powerful content recommendation systems ever built for consumer use, has been specifically documented to funnel vulnerable teenagers toward eating disorder content, self-harm content, and extreme ideological content with a consistency that the platform's own employees have reportedly flagged internally.

SURGEON GENERAL ADVISORY 2023: Up to 95% of teens 13-17 use social media, with more than one-third using it almost constantly. Teens who spend more than 3 hours daily on social media face double the risk of depression and anxiety symptoms. The average teen now spends 3.5 hours per day on these platforms already above the danger threshold.

The Slot Machine in Every Pocket

To understand why the mental health data looks the way it does, you need to understand how social media platforms are designed not as communications tools, but as attention capture and retention systems.

The variable reward mechanism the psychological principle that rewards delivered at unpredictable intervals are more compelling than rewards delivered at predictable ones is the design foundation of every major social media feed. Each scroll delivers an unpredictable mix of content: some interesting, some boring, some emotionally resonant, some forgettable. The unpredictability is not accidental. It is the same mechanism that makes slot machines more compelling than vending machines. You keep pulling the handle because you never know when the reward will come.

For teenagers whose brains are in a developmental phase that is characterized by heightened sensitivity to social rewards and social threats, and by a prefrontal cortex that has not yet fully developed the capacity to regulate impulse this mechanism is not merely compelling. It is, in the assessment of many developmental psychologists, neurologically comparable to addiction. The same reward pathways that are activated by substances are activated by social media engagement. The dopamine hit of a like, a comment, a share functions in the developing brain similarly to any other reward stimulus, and the pursuit of those rewards can become dysregulating.

The Surgeon General's advisory noted that frequent social media use could be associated with changes in parts of the brain related to emotions and learning, affecting impulse control, social behavior, emotional regulation, and sensitivity to social punishments and rewards. These are not temporary effects. They are developmental effects changes to the architecture of a brain that is still forming, potentially shaping how that brain will function for the rest of that person's life.

The platforms know this. Meta's own internal research, revealed in the 2021 whistleblower disclosures by Frances Haugen, documented that the company knew Instagram was harmful to teenage girls' body image and mental health, particularly regarding eating disorders and self-comparison. The research was presented internally. It did not change the algorithm. It did not change the design. It did not change the features. The company continued to serve the same content to the same vulnerable teenagers because the content that harmed them was also the content that kept them most engaged, and engagement was revenue.

A 2024 study published in JAMA Pediatrics offered one of the most direct demonstrations yet of the social media-mental health relationship. Teens who reduced their social media use to thirty minutes daily showed significant, measurable decreases in depression and loneliness after just three weeks. The control group, which continued normal usage, showed no improvement. Three weeks of reduced use produced measurable improvements in mental health outcomes. The intervention is not complicated. The barrier to it is not technical or medical. The barrier is the same one discussed throughout this book: the commercial interests of the companies providing the product conflict with the wellbeing of the people using it.

The Identity Experiment Run Without Consent

Adolescence has always been the developmental stage at which human beings engage in their most intense identity formation the exploration of who they are, what they believe, where they belong, and what kind of person they want to become. This process requires a degree of privacy, of experimentation, of freedom to try on identities and abandon them without consequence.

Social media does not provide this. Social media provides a permanent, public, searchable record of every aspect of the identity experiment. Every opinion posted, every affiliation signaled, every group associated with, every photo published all of it constitutes a documented record of who a teenager was at any given moment, preserved for future reference by future employers, future romantic partners, future adversaries, and the algorithmic systems that will form assessments of that person based on their digital history.

The practical consequences are already visible. College admissions offices routinely conduct social media searches of applicants. Employers screen candidates' social media profiles as a standard part of hiring processes. Law enforcement uses social media data in investigations. Political campaigns use behavioral profiles derived from social media activity to target and characterize voters. The identity experimentation that adolescence requires the trying and failing and changing your mind now generates a permanent record that follows teenagers into every consequential evaluation of their adult lives.

The psychological consequence of knowing this and teenagers increasingly do know it is a form of chronic self-censorship and performance that psychologists have documented with growing concern. Teenagers curate their social media presence with a sophistication that, in previous generations, was not required until adulthood and professional life. They are performing selfhood for an audience of hundreds from the age of twelve, under conditions of permanent record, with the knowledge that anything they post could be used against them in ways they cannot fully anticipate.

This is not healthy development. It is accelerated exposure to adult social pressures at an age when the psychological architecture to handle those pressures is not yet in place. The adults who experienced adolescence before smartphones are watching this happen from the outside and often do not fully understand what they are watching. The children experiencing it have no outside they have known nothing else.

The Predator's Paradise: When Safety Fails

No examination of children and the digital world can avoid the most acute and immediate danger: the online environment has provided unprecedented tools and access to those who seek to harm children, and the platforms that profit from children's engagement have consistently prioritized that engagement over the safety measures that would protect them.

The statistics in this area are difficult to read. They should be.

NCMEC the National Center for Missing and Exploited Children received more than 36 million reports of child sexual abuse material in 2023. Not thirty-six thousand. Not thirty-six hundred thousand. Thirty-six million. The number has grown every year for more than a decade, tracking the growth of the platforms through which this material is generated and shared. The material is not just distributed. It is being created created from images scraped from social media, from footage obtained through grooming, from AI-generated synthetic content that depicts real children and fictional ones in equivalent terms of harm.

The sharenting data discussed earlier in this chapter is directly relevant here. The photographs that parents post of their children innocent, joyful, well-intentioned are accessible to anyone with an internet connection and are used by predators both for grooming creating false rapport with children by demonstrating apparent knowledge of their lives and for the creation of what researchers call CSAM: child sexual abuse material. French child protection organization Fondation pour l'Enfance documented in a 2024 report that generative AI had become a primary tool of paedophile crime, enabling the creation of explicit synthetic imagery of real children from photographs scraped from their parents' social media accounts.

The Surgeon General's advisory explicitly flagged that adults seeking to sexually exploit children use social media platforms to target and access minors. The mechanisms are well-documented: platforms that allow direct messaging between strangers enable initial contact. Algorithmic systems that optimize for engagement with no safety weighting can create connections between vulnerable children and predatory adults. Features designed to feel intimate the streaks on Snapchat, the follower counts on Instagram, the parasocial relationships enabled by every platform lower children's defenses in ways that predators understand and exploit with deliberate sophistication.

The Senate Hearing That Changed Nothing

On January 31, 2024, the CEOs of Meta, TikTok, Snap, Discord, and X were compelled to appear before the United States Senate Judiciary Committee to account for their platforms' failures to protect children. The hearing produced moments of extraordinary moral clarity and considerable political theater.

Senator Lindsey Graham asked Mark Zuckerberg directly whether he was aware that his platform was being used to exploit children. Zuckerberg gave carefully hedged answers about the scale of the platforms and the difficulty of moderation at that scale. Senator Marsha Blackburn asked specific questions about specific documented cases of children who had been harmed through Meta platforms. Senator Josh Hawley pointed at the gallery, where a group of parents whose children had been harmed or killed as a result of social media were seated. He asked Zuckerberg to stand and face them, and to apologize directly. Zuckerberg stood. He turned to face them. He said he was sorry for everything they had been through.

The parents were photographed weeping.

Mark Zuckerberg returned to California. Meta's stock rose in subsequent weeks as quarterly earnings exceeded expectations. No legislation directly responsive to the hearing's revelations was passed in the ensuing months. The platforms remain largely unchanged.

This is not a story about one bad hearing or one failed legislative moment. It is a pattern. It is the same pattern we saw in Chapter Three with AI deepfakes. The powerful show up, express concern, and protect their business model. The parents in the gallery go home to houses where their children are still using the same platforms.

> **JANUARY 31, 2024:** The CEOs of Meta, TikTok, Snap, Discord, and X testified before the Senate Judiciary Committee on child sexual exploitation. Parents of children harmed by social media platforms attended in person. The hearing produced no comprehensive federal legislation. The platforms continue to operate substantially unchanged.

The Education System's Bargain With Big Tech

While parents and social media companies dominate public discourse about children and technology, a parallel and in some ways more insidious relationship deserves attention: the partnership between the education system and major technology companies that has put surveillance infrastructure into classrooms, onto school networks, and onto the devices that millions of children use for learning every day.

The penetration of Google and Microsoft products into American K-12 education is near-total. Google Classroom and Microsoft Teams handle the daily educational workflow of the majority of American schoolchildren. Chromebooks Google's affordable laptop, optimized for Chrome OS and deeply integrated with Google's product ecosystem are the most widely deployed student computing device in American classrooms. The education-specific versions of these products are marketed with privacy assurances, and there are legal restrictions on how student data can be used. The reality on the ground is more complicated.

What the Classroom Knows

Educational technology platforms collect extraordinary amounts of behavioral data about students. Every search conducted on a school-issued device, every document created in Google Docs, every assignment submitted through Google Classroom, every video watched on YouTube via an educational account is logged, stored, and associated with a student profile. The stated purpose of this data collection is improving educational outcomes personalizing learning, identifying students who need support, tracking progress over time.

The actual scope of what is collected and how it is used has been the subject of growing concern among privacy advocates, parents, and state attorneys general. Student surveillance software tools like GoGuardian,

Securly, and Bark monitor students' online activity on school devices and flag concerning content or behavior for administrators or parents. These tools have genuinely useful safety applications: early identification of students expressing suicidal ideation, flagging of cyberbullying, monitoring for self-harm content.

They also collect, store, and analyze the private communications of minors on a scale and with a thoroughness that would be unacceptable if applied to adults. The student who privately searches for information about their sexuality, their family's legal situation, their mental health symptoms, their political beliefs, is conducting those searches through a system that records them, associates them with their identity, and makes them accessible to institutional authorities. The privacy protections that apply to these records are imperfect, and the commercial interests of the technology companies providing these services create ongoing pressure against the strongest interpretations of those protections.

The schools that have deployed this infrastructure did so with the best intentions and, in many cases, under real financial pressure. Google and Microsoft offer educational products at deep discounts or for free. When school budgets are constrained and technology is expensive, the free or cheap solution wins, and the cost is not always fully understood at the moment of decision. The price of free technology for schools is the data generated by the students using it and that data is the foundation of profiles that will follow those students for years.

The Facial Recognition Experiment

Several American school districts have experimented with facial recognition technology for campus security scanning student faces as they enter buildings, tracking movement through hallways, identifying individuals in surveillance footage. The stated rationale is safety: faster identification of intruders, faster response to threats, improved accountability.

New York's Lockport City School District deployed a facial recognition system in 2020 over the objections of civil liberties organizations and parents. The system was ultimately shut down in 2022 after New York's education department determined it violated state law. But the incident demonstrated that the impulse to deploy biometric surveillance infrastructure in schools creating biometric databases of minors was real and would be acted on without careful legal constraints.

The child whose face is scanned at school every morning is contributing to a biometric database from the age of five. That database, properly protected today, may not remain properly protected as political conditions change, as the companies holding it are acquired or change their terms, or as law enforcement seeks access for purposes that were not contemplated when the system was installed. The child did not consent to contribute their biometric data to any database. Their parents may not have fully understood what they were consenting to. The school that deployed the system may have had the best intentions. None of this determines what happens to that data ten years from now.

The Attention Economy's Child: Manufactured Distraction and Stunted Depth

The mental health crisis documented by the Surgeon General is the most acute consequence of the attention economy's impact on children. But there is a subtler, more pervasive consequence that is harder to measure and in some ways more troubling in its long-term implications: the systematic disruption of the capacity for sustained attention, deep thought, and the kind of slow, patient engagement with difficult ideas that has historically been the foundation of human intellectual progress.

Every major platform in the attention economy is optimized for short, frequent, emotionally stimulating interactions. TikTok's default video length, already short when the platform launched, has shortened further as the algorithm learned that shorter content produced more scroll-through and therefore more overall engagement time. Instagram shifted from static posts to Stories to Reels, each format shorter, more ephemeral, more algorithmically curated for maximum engagement. YouTube's algorithm pushes shorter videos in recommendations even when users are watching longer content. Twitter reduced to X, enforcing brevity. Everywhere, the architecture rewards brevity and penalizes depth.

Children and teenagers who have grown up inside this architecture have been developing their attentional and intellectual habits within it. The attention span data is contested the oft-cited claim that the average human attention span has fallen below that of a goldfish has been criticized as methodologically dubious but the anecdotal and classroom evidence from educators is consistent and concerning: teachers at every level report increasing difficulty sustaining student attention for extended tasks, increasing resistance to long-form reading, increasing expectation of constant stimulation and immediate reward that extended intellectual work does not provide.

The educator who assigns a thirty-page reading and expects students to complete it with the same engagement that previous generations brought to it is not accommodating the attentional environment in which those students have grown up. This is not a counsel of accommodation it is an observation about the dimensions of what has been lost. The capacity to sit with a difficult text, to hold a complex argument in sustained attention, to tolerate the discomfort of not immediately understanding something these are the intellectual foundations of higher order thinking. They are skills. They are learnable. And they are being systematically undermined by platforms designed to ensure that no moment of boredom or friction occurs, because boredom and friction cause users to put down their phones.

The Radicalization Pipeline: When Algorithms Teach

The algorithmic recommendation systems that keep children and teenagers engaged on social media platforms are not politically or ideologically neutral. They are engagement-maximizing meaning they recommend content that produces the strongest emotional reactions, and the content that produces the strongest emotional reactions tends, by a consistent and documented pattern, to be the most extreme, the most provocative, the most divisive.

Research on algorithmic radicalization the process by which recommendation systems gradually introduce users to increasingly extreme content through successive recommendations has documented this pattern across multiple platforms. A user who watches moderate political commentary is recommended increasingly partisan content. A teenager who watches fitness videos is recommended increasingly extreme diet content. A young person exploring questions of identity is recommended increasingly absolutist ideological frameworks.

The JMIR study cited in this chapter's research noted that recommendation systems have resulted in the deprioritizing of content from family, friends, and accounts that users choose to follow, in favor of algorithmically curated content designed to keep users on platforms as long as possible. These curated feeds can amplify young people's exposure to extreme viewpoints misogyny, homophobia, self-harm normalization, political extremism in ways that the user did not choose and that the platform does not flag.

A thirteen-year-old whose algorithm has learned that emotionally provocative content produces longer engagement sessions is being taught by that algorithm. Not deliberately. Not maliciously. But the lesson is being taught nonetheless. And the teacher has neither a teaching credential nor the slightest interest in the student's long-term development. It has only an interest in keeping the student scrolling for one more minute.

The Future Being Written Right Now: Identity in the Age of Permanent Record

The children being raised in the current digital environment will be the first adults in human history who have no private past. Every generation before them had the ability to leave behind the worst of their younger selves to grow, change, become different people, and present that different person to a world that did not have access to who they used to be. That ability is gone for the generation growing up now, and its loss has consequences that extend far beyond the embarrassment of old photos.

The Job Application at Twenty-Two

Consider the twenty-two-year-old applying for their first serious job in 2033. Born in 2011, they have a social media history that dates to approximately 2019, when they were eight years old and their parents first created accounts for them on child-oriented platforms that transitioned, as they aged, to mainstream social media. They are twelve years old in their earliest searchable social media presence. By the time they are applying for work, they have eleven years of documented digital history.

Within that history: the opinions they held at twelve that they have since completely changed. The friendship drama of early adolescence. The political phase they went through at fifteen that no longer represents their views. The breakup they processed publicly in ways they now regret. The embarrassing content that accumulated during a difficult year in high school. The posts made in anger, in grief, in confusion, in the ordinary tumult of becoming an adult.

Employers already use social media screening as standard practice. Artificial intelligence tools for candidate screening increasingly automate this process, systematically analyzing social media histories for content that might indicate poor judgment, controversial opinions, or character concerns. The AI system doing the analysis does not apply the contextual understanding that a human reviewer might does not account for the fact that the opinion expressed at age fifteen is probably not the opinion held at twenty-two. It applies pattern recognition to digital data and produces a score or a flag that may end a candidacy before a human ever looks at the resume.

The twenty-two-year-old applying for that job has no way to erase what was posted at fifteen. In the United States, the right to be forgotten that Europe's GDPR provides in limited form does not exist at federal level for most circumstances. The posts may still be on the original platform or may have been copied, archived, or cached by systems the original poster never knew about. The digital past is permanent in a way that no previous generation's past has been.

The Credit Score That Starts in the Crib

As data analytics systems become more sophisticated and more integrated into financial decision-making, the childhood digital footprint becomes relevant not just to employment and social reputation but to financial opportunity. The behavioral data accumulated during childhood spending patterns of parents inferred from family posts, neighborhood and economic status data embedded in location tags, educational and developmental data from school systems provides raw material for the predictive analytics that increasingly govern access to credit, insurance, and financial products.

Credit scoring systems are already incorporating non-traditional data sources social media behavior, online purchase history, app usage patterns in ways that are moving rapidly ahead of regulatory frameworks. Children who grow up in the data economy with rich behavioral profiles may find that those profiles, assembled without their knowledge or consent across their entire childhoods, influence their access to financial products in adulthood.

The child whose parents documented financial stress on social media. The teenager whose school records included behavioral flags. The young adult whose early social media history, scraped and analyzed by systems they never interacted with, produces a risk score that a lender's algorithm weights against them. This is not science fiction. It is the logical extension of trends that are already observable in the financial industry's incorporation of alternative data.

The Synthetic Identity Built From Childhood Data

We examined synthetic identity fraud in Chapter Three in the context of AI deepfakes. The specific vulnerability of children to this form of fraud deserves separate emphasis here, because the mechanism exploits sharenting data in a way that parents almost never anticipate.

A child's Social Security number is issued at birth or shortly thereafter and is associated with a credit history that is, by definition, empty. An empty credit history is, paradoxically, valuable to identity fraudsters because it can be built upon a synthetic identity combining the child's real Social Security number with a fabricated name and address can be used to open credit accounts, take out loans, and build a fraudulent financial history over years or decades, with the fraud only discovered when the child reaches adulthood and attempts to use their own Social Security number for the first time.

The sharenting data that provides a child's name, birthdate, and associated personal details makes the construction of this synthetic identity easier. The fraudster does not need to breach a database to obtain a child's identifying information if that information has been shared publicly by the parents. It is there, in old Instagram posts and Facebook updates, waiting for anyone patient enough to collect it.

The child who discovers at eighteen that their credit history is already compromised that loans were taken out in their name during their childhood, that their Social Security number has been associated with debts they did not incur faces a remediation process that can take years and that begins their adult financial life with a burden that was placed on them, without their knowledge or consent, before they were old enough to walk.

What We Are Taking From Them That Cannot Be Bought Back

We have covered the data, the mental health research, the predator access, the identity risks, and the financial consequences. All of these are serious. But there is something harder to quantify that deserves its own space in this chapter something that feels less like a policy problem and more like a grief, if you sit with it long enough.

We are taking childhood from children.

Not the whole of it. Not deliberately. Not with malice. But the evidence is consistent and it points in one direction: the children growing up inside the attention economy are spending less time in the unstructured, unsupervised, self-directed exploration that developmental psychologists have long understood to be fundamental to healthy human development. They are spending less time outdoors. Less time in physical play. Less time in the slow, patient, face-to-face social negotiation that teaches empathy and conflict resolution and the tolerance of difference. Less time bored and boredom, uncomfortable as it is, is the condition in which creativity and self-directed imagination emerge.

They are spending that time scrolling. Being fed. Being sold. Being analyzed. Being nudged toward content that keeps them engaged for another minute, another hour, another session, because another session means another serving of advertisements to a developing brain that has not yet built the defenses to evaluate what it is receiving.

Jonathan Haidt, a social psychologist at NYU and one of the leading researchers on social media and youth mental health, argues in his 2024 book The Anxious Generation that the smartphone has fundamentally restructured childhood in ways that are producing a generation defined by anxiety, loneliness, and an inability to tolerate the discomfort that used to be the ordinary texture of growing up. He is careful about causation. He is not careful about urgency. The subtitle of his book is How the Great Rewiring of Childhood is Causing an Epidemic of Mental Illness. That is not a hedged claim.

We are also taking from children something that no algorithm can manufacture and no platform can replicate: the experience of being known in their full, unmediated, private complexity by people who love them, in spaces where they do not have to perform, where they are not being watched and recorded and analyzed, where they are simply, privately, imperfectly, beautifully themselves.

That experience of private selfhood, of unobserved development, of a childhood that belongs to the child is not a luxury from a more innocent time. It is a developmental necessity, and we are allowing it to be stripped from the children in our care, one algorithm-optimized minute at a time.

The Horizon: What Is Already Being Built for a Better Future

We have spent most of this chapter in the dark. That was deliberate. The stakes are high enough that false comfort would be a disservice. You needed to see clearly what is happening before you could properly appreciate why what comes next matters so much.

Because here is the thing about children that should fill every parent, grandparent, teacher, and citizen with something that is not despair but determination: children are not passive. The generation growing up inside the attention economy is also the generation that is most acutely aware of its dynamics, most vocal about its harms, and most hungry for a different relationship with technology if adults will do the work of building one.

And increasingly, they are.

The movement toward digital rights for children toward legal frameworks that treat the data of minors as requiring the highest level of protection, that recognize the developing brain as deserving special consideration rather than special exploitation, that hold platforms accountable for the environments they create and the harms those environments cause is real, documented, and accelerating.

This is the hopeful part. It is grounded in specifics, not wishful thinking. Let us go through it carefully, because it deserves the same careful attention we gave to the crisis.

Legislative Momentum: The Laws Being Written

The Kids Online Safety Act KOSA passed the U.S. Senate with a 91-3 vote in July 2024, the strongest congressional bipartisan showing on child protection legislation in decades. The bill would require social media platforms to implement the strongest privacy settings by default for users under the age of eighteen, restrict platforms from promoting content to minors that could cause harm, and give minors the ability to opt out of algorithmic recommendation systems entirely to see content from accounts they choose to follow rather than what the algorithm decides to show them.

The Children and Teens' Online Privacy Protection Act COPPA 2.0 passed the Senate in the same session, expanding the original 1998 law's protections to cover users up to sixteen (rather than thirteen) and creating a Youth Marketing Prohibition that would prevent platforms from targeting advertising to minors based on behavioral data.

These bills represented the most significant movement toward comprehensive federal child online safety legislation in twenty-five years. Their passage through the Senate with overwhelming bipartisan support demonstrated that child protection is one of the issues on which the polarized American political system can find genuine common ground. The bills faced a more uncertain path in the House, where lobbying from technology companies exerted significant pressure.

But the direction is clear. The political will exists. The public mandate is not ambiguous. The question is whether the organized pressure of citizens who understand what is at stake will be sufficient to push this legislation over the finish line.

> **SENATE VOTE:** The Kids Online Safety Act passed the U.S. Senate 91-3 in July 2024 one of the most bipartisan votes in recent congressional history. The bill would require default privacy settings for minors and give teens the ability to opt out of algorithmic recommendation entirely. This is what sufficient public pressure produces.

The Right to Be Forgotten: Europe Shows the Way

The European Union's General Data Protection Regulation includes, under Article 17, a right to erasure the right for any individual to request the deletion of their personal data from digital systems. For children, the GDPR's protections are stronger still: data collected from minors requires specific consent processes, and the right to erasure is explicitly supported for data collected during childhood.

France has gone further. In June 2024, France became the first country in the world to ban smartphones for children under eleven and restrict access for those under fifteen, requiring phones used in schools to have parental controls enabled. The French government framed this not as technophobia but as the protection of childhood recognizing that the rights of children require active protection, not passive hope that the market will sort it out.

Australia passed comprehensive age verification legislation in November 2024, banning children under sixteen from creating social media accounts a prohibition backed by enforcement mechanisms that place responsibility on the platforms rather than on children and parents to navigate alone. Prime Minister Anthony Albanese stated plainly that the social media companies needed to be held responsible for the damage they cause, and that protecting children was not a partisan issue but a basic obligation of government.

These are not perfect solutions. Age verification systems have technical limitations and privacy implications of their own. The French phone ban produces implementation challenges. The Australian law faces constitutional questions. But they represent something critically important: governments making a deliberate choice that the commercial interests of technology companies do not automatically outweigh the developmental rights of children, and that the state has both the authority and the obligation to impose that priority.

The United States has not yet made that choice at the federal level. But every law passed elsewhere is evidence that the choice can be made, and that the civilization-ending consequences that technology industry lobbyists predict from any meaningful regulation do not in fact materialize.

AUSTRALIA 2024: Australia banned children under 16 from social media with enforcement obligations on the platforms themselves not on children or parents. France banned smartphones for children under 11. These governments chose children over corporate revenue. That choice is available to every democracy.

The Parent Movement: Organizing for Our Children

One of the most heartening developments in the fight for children's digital rights is the emergence of organized parent advocacy that has proven capable of moving both legislatures and platforms in ways that individual parents acting alone cannot.

Wait Until 8th a pledge movement asking parents to commit to not giving their child a smartphone until at least eighth grade has gathered hundreds of thousands of signatories. The movement's power is specifically in its collective dimension: individual parents who want to delay smartphone adoption often face social pressure from their children, who point out that everyone else has a phone. When large numbers of parents in a school or community make the commitment together, the social pressure disappears because the peer environment changes.

The 1000 Hours Outside movement encouraging parents to prioritize unstructured outdoor time for children has similarly attracted enormous interest among parents who feel intuitively that something has gone wrong with the balance of their children's lives and are looking for community support in making a different choice.

Jonathan Haidt's research and advocacy, particularly through his book The Anxious Generation and the organization Let Grow, has provided parents, educators, and policymakers with a research-grounded framework for understanding what is happening and what can be done. His specific recommendations phone-free schools, no smartphones before high school, delaying social media until sixteen, more unsupervised outdoor play are not technophobic. They are developmentally grounded, practically achievable, and already being implemented in schools and communities that have decided not to wait for platforms or governments to act.

Phone-Free Schools: The Revolution Already Happening

Perhaps the most concrete and immediately actionable change happening right now in children's digital lives is the rapid spread of phone-free school policies and the data on their effects is providing some of the most encouraging evidence in the entire field.

Schools that have implemented strict phone-free policies not phone-in-pocket policies, but genuine phone-away-in-locker-or-pouch policies consistently report the same outcomes: reduced bullying, improved focus, increased in-person social interaction during breaks, and measurable improvements in academic engagement. The students themselves, after an initial adjustment period, frequently report that they prefer the phone-free environment.

A 2024 study of phone bans in schools found that the benefits were greatest for the most vulnerable students those experiencing mental health difficulties and those from lower-income backgrounds supporting the view that the addictive and comparison-generating properties of social media disproportionately harm those who are already struggling. Phone-free schools do not solve the social media problem at home. But they create, for six or seven hours of every school day, an environment in which children can develop social skills, focus, and the experience of sustained attention without the constant interruption of a device designed to maximize that interruption.

By early 2025, more than twenty American states had passed legislation either mandating or strongly encouraging phone restrictions in schools. The movement has broad political support, cutting across party lines in the way that genuine child welfare issues often do when they are framed clearly. This is a policy change happening now, producing measurable benefits now, driven by parents and educators and legislators who decided that the evidence was sufficient to act.

> **PHONE-FREE SCHOOLS:** Over 20 U.S. states have passed legislation restricting phones in schools by early 2025. Studies consistently find reduced bullying, improved focus, and increased social connection. The benefits are greatest for the most vulnerable students. This is a change that works, already happening, driven by people who refused to wait.

The World We Could Still Build: A Vision for Children Born Ten Years From Now

Let us do something that books about digital threats rarely do. Let us describe, specifically and concretely, what a better world looks like not as utopian fantasy, but as the practical consequence of the legislative, cultural, and technological choices that are already being made in some places and that could be made everywhere.

Imagine a child born in 2035.

Their parents, educated in the privacy implications of sharenting, have made a deliberate choice to keep their child's image off public social media. The photographs taken in those first weeks exist in a private, encrypted family album shared only with people who specifically matter. The baby's face is not in any commercial database. Their birth announcement did not include their full name, their date of birth, and their precise location. Their digital footprint, at birth, is effectively zero.

The school they attend in 2040, at age five, operates under a federal digital privacy framework that prohibits the commercial use of student data. The educational technology used in their classroom is privacy-preserving by design the minimum data necessary for educational purposes is collected, used for that purpose, and deleted on a defined schedule. There is no hidden behavioral profiling. There is no advertising targeting. There is no commercial surveillance of what they search for when they are trying to understand the world.

Their school is phone-free. They learn to read, to calculate, to reason, to imagine in an environment that is not competing with an infinite scroll of algorithmically optimized stimulation. They develop the capacity for sustained attention because they have been given the conditions in which that capacity develops. They play outside. They are bored sometimes. They figure out what to do with the boredom.

When they are twelve, the social media platforms that exist have been required to implement the strongest possible privacy settings by default, to abstain from behavioral advertising targeting minors, to provide algorithmic transparency about what content is being recommended and why, and to make it genuinely easy to use chronological feeds rather than engagement-optimized ones. The algorithm does not radicalize them because platforms have been held legally accountable for the content their recommendation systems amplify. The platforms exist to connect people, not to maximize time on site by any means necessary.

When they are sixteen and ready to engage more fully with the digital world, they have been taught in school, at home, through a culture that takes digital literacy seriously how the attention economy works, how behavioral advertising works, how data is collected and used and sold, how to recognize manipulation and protect against it. They approach the digital world not as naive subjects of its systems but as informed participants who understand the terms on which they are engaging.

When they are eighteen, they can review the digital data that has been collected about them during their childhood and request its deletion. The right to a digital clean slate at adulthood is a legal right, not a technicality. They enter adulthood with a past that belongs to them one they can choose to share or not share, one that reflects the person they have become rather than a permanent record of every awkward, confused, experimental moment of childhood.

When they apply for their first job at twenty-two, employers cannot conduct algorithmic screening of their childhood social media history because that history is protected and because the most privacy-invasive data collection practices have been restricted by law. They are assessed on their qualifications, their skills, and their character as they present themselves not on a behavioral profile assembled by systems they never knew existed.

Is this utopia? No. It is a description of a world that incorporates the best policies currently being implemented in various countries and communities that takes the most protective elements of European digital rights law, Australian social media age restrictions, American phone-free school policies, and emerging federal child online safety legislation, and extends them consistently and coherently.

Every element of this vision already exists somewhere. Every policy described is already law in some jurisdiction. Every technology mentioned is technically feasible. Nothing in this future requires a breakthrough. It requires choices. It requires political will. It requires citizens who understand what is at stake and demand that their governments act.

What You Can Do Right Now: Starting Today

The future described above does not arrive on its own. It arrives because people who care about children decide to act in their homes, in their communities, and in the political arena. Here is where to start.

In Your Home

Pause before you post. The single most impactful thing parents can do for their young children's digital futures is to develop a conscious, deliberate approach to sharing information about them online. This does not mean never sharing. It means asking, before each post: Would this child, as an adult, be glad this exists? Does it identify them by name in a public forum? Does it reveal their location? Does it contain content they might find embarrassing or harmful? It means setting social media accounts to private when sharing family photos. It means treating your child's digital footprint as a resource that belongs to them, not to your social media audience.

Delay smartphones and social media. The evidence for delaying smartphone adoption until at least eighth grade is strong, consistent, and growing. Every year of delay is a year of development that occurs outside the addictive architecture of the attention economy. Every year of delay is a year in which your child's brain forms in conditions more conducive to the sustained attention, deep social connection, and genuine selfhood that human flourishing requires. This is hard. It requires saying no to social pressure. It is worth it.

Make your home phone-free during certain times. Family dinner without phones. An hour before bed without screens. The specific implementation matters less than the principle: that there are times and spaces in your home where the devices do not compete with human presence. Research on the effect of phones on family interaction is unambiguous even the mere presence of a phone on a dinner table, face down and silent, measurably reduces the quality of conversation. Put the phone in another room. See what happens.

Talk to your children about how the attention economy works. Not in a way that frightens them or makes them feel surveilled by their own parents. In a way that equips them. Teenagers who understand that the like button was designed to trigger dopamine are in a meaningfully different position than teenagers who experience it as natural social validation. The teenager who knows that the algorithm is showing them inflammatory content because engagement drives revenue is in a better position to evaluate what they are consuming than the teenager who experiences it as spontaneous discovery.

In Your Community

Join or start a Wait Until 8th pledge in your school community. The power of the collective commitment cannot be overstated. When enough parents in a school community make the same decision, the social pressure on individual families disappears. Find other parents who share your concerns. Build the coalition that makes the different choice feel normal rather than eccentric.

Advocate for phone-free school policies. If your children's school does not have a strong phone policy, ask why. Bring the research to your parent-teacher organization. Bring the data on reduced bullying, improved focus, and increased social connection. Most school administrators are aware of the problem and many are looking for parent support to implement policies that they know will face initial resistance from students.

Support your local library, bookstore, and the spaces that model the alternative to the attention economy. The child who grows up going to the library, who is read to, who lives in a home where books and sustained attention are valued, is being given something that the attention economy is trying to take from them. These choices are not nostalgic. They are resistance. They model a different relationship with time, attention, and the consumption of ideas.

In the Political Arena

The Kids Online Safety Act, COPPA 2.0, and the NO FAKES Act's provisions relevant to children are all pending, stalled, or inadequately enforced. They all need champions constituents who contact their representatives, who vote based on these issues, who make clear that child online safety is not a niche concern but a mainstream priority.

The Federal Trade Commission has jurisdiction over unfair and deceptive practices in the children's digital market and has used that jurisdiction with varying degrees of aggressiveness across administrations. The FTC needs to hear from citizens who want strong enforcement of existing child privacy protections and strong action against violations.

Support candidates and elected officials who have demonstrated genuine commitment to child online safety not performative commitment, shown in one hearing and forgotten, but sustained, legislative commitment demonstrated by voting records and willingness to stand against technology industry lobbying.

The political will to protect children online exists in the American public in overwhelming proportions. Polling consistently shows broad, bipartisan support for stronger regulation of social media companies' impact on children. That public will is not yet being fully translated into legislative action because the organized pressure of the technology industry is greater, at this moment, than the organized pressure of citizens who care about children's wellbeing. That is a solvable problem. It is solved by people who care about children deciding to become as organized, as persistent, and as politically engaged as the industry whose interests oppose them.

The Letter You Are Writing to the Future

There is a specific kind of moral clarity that comes when you think about children not statistics, not policy categories, but actual children. The specific child in your life whose face you know, whose laugh you recognize, whose fears and hopes you care about in a way that is not abstract.

That child is going to grow up. They are going to be twenty-two, and then forty, and then sixty, and they are going to look back on the world they inherited from the people who were adults when the decisions that shaped it were being made. They are going to know what those adults knew. They are going to know what the Surgeon General said in 2023, what the Senate hearing revealed in January 2024, what the research on sharenting has documented, what the Volt Typhoon intrusions demonstrated, what the deepfake crisis is already doing to vulnerable people.

They are going to know all of it, because it is documented, because we cannot hide it. And they are going to ask: what did you do?

That question is not punitive. It is not a condemnation of anyone who did not understand these issues before reading this book or before the research made the scale of the problem undeniable. It is the entirely reasonable question that every generation asks of the one before it: when you saw the problem, when the evidence was clear, when the tools to act were available what did you do?

The beautiful and difficult truth is that the answer is not yet written. The choices that will determine what that adult looks back on are being made right now. The legislation is pending. The phone policies are being debated. The digital rights frameworks are being drafted. The cultural norms around sharenting are shifting. The parent movements are organizing. The researchers are publishing. The courts are beginning to rule.

Everything is still possible. The child born today does not have to inherit the worst version of the digital world that the surveillance economy and the attention economy and the foreign state hackers and the deepfake fraudsters have been building. They can inherit a version that was shaped by people who understood what was at stake and refused to accept that the current trajectory was inevitable.

That refusal begins with knowledge. You now have it. It continues with action. The next chapter will give you a comprehensive framework for that action the specific, practical, political, and personal steps that translate understanding into change.

But first: take a moment. Think about the child in your life. Think about the world they are already living in, and the world they will live in, and the world you would choose for them if choosing were simple and the choice were yours alone.

Now understand that choosing is not simple. But the choice is yours.

Not yours alone. Never alone. But yours. Your voice, your vote, your choices at home, your advocacy in your community, your conversations with other parents and teachers and neighbors and elected officials all of it aggregates into the political and cultural environment that will either protect children or fail to protect them.

The generation being raised right now is the first to grow up fully inside the digital world. They are also the first generation that could grow up inside a digital world that was built to protect them rather than exploit them if we make the choices, today, to build it that way.

They are worth it.

We already knew that.

Now we have to act like it.

* * *

END OF CHAPTER FIVE

Continue to Chapter Six: Taking Back Control

CHAPTER SIX

Taking Back Control:

The Complete Playbook for Protecting Yourself, Your Family, Your Wealth, and Your Humanity in the Digital Age

---

"The most common way people give up their power is by thinking they don't have any." Alice Walker

You Have Been Paying Attention. Now Act.

You have read five chapters of uncomfortable truth. Corporate surveillance that monetizes your most intimate behavioral data. State-sponsored cyber armies burrowed inside the infrastructure that keeps your city running. Artificial intelligence that can steal your face and voice in minutes. A financial system quietly engineering the end of ordinary ownership. And a generation of children growing up as the first fully surveilled humans in history.

If you have felt the weight of it the slow accumulation of dread that comes from understanding the full scope of what is happening that feeling is appropriate. It is the correct emotional response to accurate information. Alarm is not hysteria. It is intelligence.

But alarm is only the first step. Alarm that does not produce action is just anxiety with better sources. And this book was never intended to produce paralysis. It was intended to produce the specific, clear-eyed, determined kind of anger that burns steadily and consistently enough to drive change in your home, in your habits, in your community, and in the political arena where the structural decisions get made.

This final chapter is your playbook. It is organized around four domains of action: protecting yourself digitally, protecting your financial autonomy, protecting your children, and building and defending the irreplaceable human capabilities that no algorithm will ever take from you. Each section gives you specific, actionable, current tools and approaches not vague encouragement, but the exact steps that turn awareness into armor.

You are not powerless. The people who want you to believe you are powerless have a financial and political interest in your believing it. The surveillance economy depends on your passive participation. The attention economy depends on your distraction. The political machinery that has consistently failed to protect ordinary people depends on your disengagement.

The moment you stop being passive, distracted, and disengaged is the moment their leverage over you diminishes. So let us begin.

Part One: The Digital Fortress Protecting Your Privacy and Communications

Digital security is not a single product you purchase or a single setting you toggle. It is a posture a set of habits, tools, and decisions that compound over time into genuine protection. The goal is not perfect security, which does not exist. The goal is making yourself a harder target than the vast majority of people, such that the low-effort automated attacks that claim the most victims cannot claim you, and the higher-effort targeted attacks face enough friction to be deterred or detected.

Think of it in layers. Each layer you add makes you meaningfully safer. No single layer is sufficient. All of them together create something powerful.

Layer One: Your Communications Lock the Most Personal First

Your communications your calls, your messages, your emails are the most intimate layer of your digital life and also, as Salt Typhoon demonstrated with devastating clarity, among the most vulnerable. Standard phone calls and SMS text messages travel through telecommunications infrastructure that has been confirmed compromised by foreign state actors. CISA has explicitly and publicly recommended that Americans switch to encrypted communications platforms. This is the first change to make.

Signal is the gold standard. It is free, open-source meaning independent security researchers have verified its code available on iPhone and Android, and protects both messages and voice calls with end-to-end encryption so strong that even Signal's own servers cannot read your communications. The encryption protocol that Signal developed has been so thoroughly validated that it is used by

WhatsApp, iMessage, and other major platforms. The difference is that Signal collects virtually no metadata about your communications not who you talked to, not when, not for how long.

Make Signal your default for conversations with anyone who matters. Your family. Your close friends. Anyone with whom you discuss anything sensitive financial matters, health information, professional concerns, anything you would not want read by a stranger. The switch takes ten minutes and the app is free.

For email, consider moving sensitive communications to ProtonMail, which offers end-to-end encrypted email and is headquartered in Switzerland under some of the strongest privacy laws in the world. ProtonMail's free tier is functional for most individual users. The full Proton ecosystem which includes Proton VPN, Proton Drive for encrypted cloud storage, and Proton Pass for passwords offers one of the most comprehensive privacy-respecting alternatives to Google's product suite currently available.

SIGNAL: Free, open-source, end-to-end encrypted messaging and calling. Available on iPhone and Android. Recommended explicitly by CISA following Salt Typhoon. Switch today at signal.org. It takes ten minutes.

PROTONMAIL: End-to-end encrypted email, Swiss jurisdiction. Free tier available at proton.me. The full Proton Unlimited plan adds VPN, encrypted storage, password manager, and calendar in a single privacy-first ecosystem.

Layer Two: Your Passwords The Credential Crisis

Credential theft the theft of usernames and passwords is the single most common entry point in virtually every category of cyberattack, from personal account takeovers to the corporate network intrusions that Volt Typhoon used to penetrate critical infrastructure. The solution has been known for years and is still massively underdeployed: unique, strong passwords for every account, managed by a dedicated password manager.

The statistics on this are both encouraging and sobering. Research from 2025 finds that only about 36% of American adults roughly 94 million people use a password manager. Among those who do, the identity theft rate is 17%. Among those who do not, the rate is 32%. The protection gap is almost exactly double. The tool is available, the evidence is clear, and nearly two-thirds of the country remains unprotected.

For most users, Bitwarden is the best starting recommendation. It is free for individuals, open-source with independently audited code, easy to use across all devices and browsers, and stores your encrypted vault in a way that even Bitwarden's servers cannot access. If you want the premium tier for about ten dollars a year, it adds encrypted file storage, two-factor authentication support, and vault health reports.

For users who want a complete privacy ecosystem, Proton Pass integrates seamlessly with ProtonMail and Proton VPN. For those who want polished design and strong family sharing features, 1Password has long been the premium choice at roughly three dollars per month. NordPass, from the team behind NordVPN, uses XChaCha20 encryption even more modern than the AES-256 standard used by most competitors and is particularly well-suited for users who want simplicity.

The master password you choose for your password manager is the most important password in your digital life. Make it a passphrase four or more unrelated words strung together rather than a complex string of characters. Passphrases are both stronger against brute-force attacks and easier to remember. Write it down on paper and store it somewhere physically secure. Do not store it digitally anywhere.

BITWARDEN: Free, open-source password manager, independently audited. The best starting point for most users. bitwarden.com free for individuals, $10/year for premium.

PROTON PASS: Open-source, Swiss-jurisdiction password manager with email alias generation. Free tier available; integrates with full Proton privacy ecosystem. proton.me/pass

Layer Three: Two-Factor Authentication The Second Lock

A password, no matter how strong, is a single point of failure. If it is stolen through a phishing attack, a data breach, or malware, your account is compromised regardless of the password's complexity. Two-factor authentication requiring a second verification step beyond the password eliminates this single point of failure.

Enable two-factor authentication on every account that offers it, starting with your email, your financial accounts, and your social media. Use an authenticator app Google Authenticator, Authy, or the built-in authenticator in your password manager rather than SMS-based verification codes. As Salt Typhoon demonstrated, SMS messages can be intercepted through compromised telecom infrastructure. Authenticator app codes exist only on your device and cannot be intercepted in transit.

Hardware security keys physical devices like the YubiKey that you plug into your computer or tap to your phone provide the highest level of two-factor protection available. For accounts that support them, particularly for high-value targets like email, financial accounts, and work systems, a hardware key is the gold standard. They cost between twenty-five and sixty dollars and provide near-impenetrable protection against phishing and credential theft.

Layer Four: Your Browser and Search Stop the Tracking at the Source

The browser you use and the search engine you use are among the most significant determinants of your digital privacy posture, and switching them is among the highest-leverage, lowest-effort changes you can make.

Google Chrome is, by design, a data collection platform. It is built by a company whose entire revenue model depends on behavioral data, and its default settings are calibrated to maximize that data collection. The Brave browser is the strongest privacy-first alternative for most users it blocks third-party trackers and advertisements by default, meaning the tracking infrastructure described in Chapter One is disabled without any additional configuration required. Firefox, with the uBlock Origin extension installed, provides comparable protection with more flexibility for users who want granular control.

For search, replace Google with DuckDuckGo or Brave Search. Both provide genuinely useful search results without building behavioral profiles of their users. DuckDuckGo has been the privacy search standard for years. Brave Search uses its own independent index it does not rely on Google or Bing data making it one of the most genuinely independent search alternatives available.

Install uBlock Origin in whatever browser you use. It is free, open-source, and blocks the overwhelming majority of tracking scripts and advertising networks that follow you across the web. The researchers who first documented the third-party tracker ecosystem the invisible layer of surveillance described in Chapter One consistently cite uBlock Origin as the most effective single tool for disrupting it.

BRAVE BROWSER: Blocks ads and trackers by default, no configuration required. brave.com available on desktop and mobile. Switch your default browser today.

DUCKDUCKGO: Private search engine, no behavioral profiling. duckduckgo.com set as your default search engine in any browser in under two minutes.

uBLOCK ORIGIN: Free, open-source ad and tracker blocker. Install from your browser's extension store. The single most effective tool for disrupting third-party tracking.

Layer Five: Your VPN Encrypting the Connection

A Virtual Private Network encrypts your internet traffic between your device and the VPN server, preventing your Internet Service Provider itself a significant data collector and any network-level surveillance from seeing what sites you visit and what data you transmit. It also masks your IP address, making it harder for websites and tracking networks to identify and locate you.

VPNs are not magic privacy tools. They shift trust from your ISP to the VPN provider, meaning the VPN provider's trustworthiness is critical. Choose a VPN with a verified no-logging policy independently audited, not just self-attested and headquarters in a jurisdiction with strong privacy laws.

Mullvad VPN is the most privacy-committed option currently available. It accepts cash payment, does not require an account with any personal information, has passed multiple independent audits of its no-logging claims, and offers a DAITA feature Defense Against AI-guided Traffic Analysis that protects against the most sophisticated traffic analysis techniques. It costs about five dollars per month.

ProtonVPN, from the Swiss Proton privacy ecosystem, is the strongest alternative, particularly for users who want integration with ProtonMail and the broader Proton suite. NordVPN and Surfshark are well-audited consumer options at competitive price points. ExpressVPN offers consistently fast speeds for users for whom performance is a priority.

Use your VPN consistently on public Wi-Fi coffee shops, airports, hotels where network surveillance is trivially easy. For home use, the benefit is primarily protection from ISP data collection, which is meaningful but less urgent.

MULLVAD VPN: The most privacy-committed VPN available. No account required, accepts cash, independently audited no-logging policy. mullvad.net approximately $5/ month.

Layer Six: Data Brokers Demanding Your Name Back

The data broker industry that maintains comprehensive profiles on virtually every American adult cannot be defeated entirely by individual action. But it can be significantly disrupted, and the legal landscape is shifting in ways that give individuals more tools to demand their data back.

DeleteMe is a subscription service that continuously opts you out of major data broker databases on your behalf. Because data brokers continuously re-accumulate data from new sources, opt-outs are not permanent they require ongoing monitoring and repetition. DeleteMe automates this process for approximately one hundred dollars per year. For individuals with particular privacy concerns executives, journalists, domestic violence survivors, anyone being stalked or harassed this is one of the highest-value privacy investments available.

For those who prefer the manual approach, the major data brokers Spokeo, BeenVerified, Whitepages, Intelius, PeopleFinder, PeekYou, Acxiom, LexisNexis each maintain opt-out procedures accessible through their websites. The process is time-consuming and must be repeated periodically, but it is free. The Privacy Rights Clearinghouse and EFF's Surveillance Self-Defense guide maintain updated lists of data broker opt-out procedures.

California residents have specific legal rights under the California Consumer Privacy Act to request deletion of their personal data from businesses that collect it. Similar rights exist under Virginia's Consumer Data Protection Act, Colorado's Privacy Act, and other state privacy laws. If you live in a state with comprehensive privacy legislation, understanding and exercising your rights under that law is meaningful.

DELETEME: Automated data broker opt-out service. Removes and monitors your personal information from major data broker databases. joindeleteme.com approximately $100/year for individuals.

Part Two: Protecting Against Deepfakes and AI Impersonation

The AI impersonation threat described in Chapter Three requires a different set of defenses than traditional cybersecurity threats, because the primary attack vector is not your devices or your accounts it is your identity, your relationships, and the trust that exists between you and the people who love you. The defenses are partly technical and partly human, and the human defenses are in many ways more important.

The Family Safe Word: The Most Important Defense You Will Set Up Today

Do this today. Before you finish reading this chapter. Establish a family code word a specific, memorable, completely random word or short phrase that has no connection to anything associated with your family and communicate it to every member of your immediate family. It must be known only to family members. It must be required on any call from a family member claiming to be in distress, in danger, or in need of financial help.

The word is the verification. If the caller cannot provide it, the call is fraudulent regardless of how perfectly it replicates the voice of a family member. Voice cloning can replicate timbre, accent, cadence, and emotional texture. It cannot know a privately held code word.

Supplement this with a verification rule: if you receive a distress call from a family member, hang up and call them back directly on a number you have independently verified a number from your contacts, not a number provided during the emergency call. If they answer, the emergency was fraudulent. If they do not answer, call another family member or trusted contact before taking any financial action.

The urgency that characterizes these scams the pressure to act immediately, before you can think or verify is manufactured. It is a manipulation technique. Real emergencies tolerate a five-minute delay for verification. Scams do not. Any caller who insists that you cannot take five minutes to verify their identity before wiring money is not your grandchild. They are a criminal.

> **DO THIS NOW:** Establish a family code word. Text or call every family member with it today. Make it a rule: no code word, no action. This single step defeats the most common and devastating deepfake scam targeting ordinary families.

Investment Skepticism: The Elon Musk Rule

Apply an absolute rule to any investment opportunity promoted in a video featuring a famous person: assume it is fraudulent until proven otherwise through multiple independent verification steps conducted over at least forty-eight hours.

No legitimate investment opportunity disappears in forty-eight hours. No real financial product requires immediate action based on a social media video. The combination of a famous face, an extraordinary return promise, and artificial urgency is the exact fingerprint of deepfake investment fraud. When you see all three elements together, you are not looking at an opportunity. You are looking at a trap.

Verify through official channels only. The official website of the company, reached by typing the URL directly rather than clicking a link. The official customer service number from the company's official website. A licensed financial advisor whose credentials you can independently verify. Never through a link, phone number, or contact information provided in the video or in a message that arrived with the video.

Protecting Your Voice and Image Online

Your voice samples and your photographs are the raw material from which deepfakes of you are constructed. You cannot eliminate this exposure if you use social media, but you can minimize it deliberately.

Set your social media accounts to private. This does not eliminate risk your followers can screenshot and share but it removes your content from public-facing data collection. Review the audience settings on your historical posts. Delete content that you no longer want public, understanding that cached copies may persist in archives but removing original content still meaningfully reduces accessibility.

Be conscious of video content specifically. Voice cloning requires audio samples. The more audio you have publicly posted particularly clear, uninterrupted speech the richer the source material for cloning. This is not a reason to disappear from social media. It is a reason to be deliberate about what you post and who can access it.

For professionals whose voice and image are public assets journalists, public figures, executives, speakers consider registering your voice with voice authentication services as they develop. Several companies are building voice signature registries specifically to allow identity verification and provide legal standing for unauthorized use claims. The technology is early-stage but moving quickly.

Verifying What You See and Hear

In the age of deepfakes, visual and audio verification is no longer sufficient as a standalone authentication method. Develop a habit of second-channel verification for any high-stakes request.

For unexpected requests financial transfers, sensitive information sharing, unusual actions requested via video call require out-of-band confirmation. Confirm via a different communication channel. Send an email to a known address in addition to responding to the call. Call a secondary contact to verify. The Arup finance employee who lost twenty-five million dollars to deepfake executives had no protocol requiring this. You can establish one right now, in your personal life and in your professional one.

Use AI detection tools as a supplementary layer, understanding their limitations. Tools like Hive Moderation, Sensity AI, and various browser extensions attempt to detect AI-generated content. They are imperfect detection technology consistently lags behind generation technology but they can flag obviously synthetic content and add friction to attacks that rely on speed and confidence.

Part Three: Protecting Your Financial Autonomy

The financial threats described in Chapters One and Four the surveillance capitalism that monetizes your financial behavior, the ownership crisis, the debt architecture, the CBDC surveillance potential require both defensive and affirmative responses. Defense means reducing your exposure to financial surveillance and manipulation. Affirmative response means deliberately building the forms of ownership and autonomy that the current system is eroding.

Understanding and Defending Your Financial Data

Your financial data your transaction history, your credit behavior, your insurance history, your income patterns is harvested and sold by a data broker ecosystem that operates with almost no accountability to you. Begin by understanding what is out there.

Request your free annual credit reports from all three major bureaus Equifax, TransUnion, and Experian through AnnualCreditReport.com, the only federally authorized source. Review them carefully for accounts you did not open, inquiries you did not authorize, and addresses you have never lived at. These are the primary indicators of identity theft and synthetic identity fraud. Place a credit freeze with all three bureaus it is free, it prevents new credit accounts from being opened in your name without your explicit unfreezing action, and it is the single most effective protection against new-account identity fraud.

Opt out of pre-screened credit and insurance offers through OptOutPrescreen.com the official consumer opt-out site operated by the major credit bureaus. These offers use your credit data to target you and represent both a privacy exposure and a scam risk.

Be deliberate about which financial apps have access to your bank account data. Open banking has created a category of apps budgeting tools, investment trackers, loan applications that require you to share your bank credentials or connect your account through data aggregators like Plaid. The convenience is real. The privacy implications are significant. Every app with access to your transaction data is a data collection point. Limit this access to apps you genuinely need and trust.

> **CREDIT FREEZE:** Free at all three bureaus. Prevents new credit accounts from being opened in your name. The most effective protection against identity theft. Freeze immediately at equifax.com, transunion.com, and experian.com.

The Ownership Imperative: Building What They Want You to Rent

Every chapter of this book has circled back to the same principle: ownership is power. The forces arrayed against ordinary ownership the subscription economy, the rental housing market, the debt architecture are powerful but not omnipotent. Deliberate, sustained commitment to building ownership rather than renting access is both personally protective and politically meaningful.

Homeownership remains the single most powerful wealth-building tool available to most American families despite the affordability challenges documented in Chapter Four. If you do not own a home and homeownership is financially feasible for you even at the margins of feasibility the long-term wealth accumulation argument for ownership over rentership is robust across virtually every market and time horizon. The mortgage payment that builds equity is categorically different from the rent payment that builds your landlord's equity, even when the monthly numbers are similar.

For those for whom homeownership is currently out of reach, the affirmative action is building the financial foundation that makes it possible eliminating high-interest debt, building a down payment, maintaining and improving credit. These are long-term plays that require discipline. They are also the specific financial behaviors that the debt architecture is designed to undermine. Every credit card balance paid off is a refusal to participate in the wealth extraction system on its preferred terms.

Beyond housing, build ownership in the forms that are accessible to you. Physical assets precious metals, tools, land if accessible retain value across financial system disruptions in ways that digital assets and accounts do not. A small allocation of savings to physical gold or silver is not paranoid hoarding. It is diversification against the specific risk of currency instability and digital financial system disruption that the CBDC debate has made more salient.

Invest in skills and tools that produce income and value independent of employment the same principle applies in the financial domain as in the human capabilities domain we will address shortly. A person who can grow food, repair things, build things, create things with their hands has a form of economic resilience that no algorithm can disrupt and no subscription model can capture.

Cash: The Privacy Tool That Fits in Your Pocket

Use cash for purchases where privacy matters and where cash is accepted. This is not a counsel of wholesale return to cash-only living the convenience of digital payments is genuine and the practical barriers to all-cash life are real. It is a counsel of deliberate, selective cash use as a privacy practice and as a political statement.

Every cash transaction is a transaction without a data record. Every cash transaction is a small act of financial privacy in a system designed to eliminate it. Every cash transaction supports the continued acceptance of cash by businesses, which matters because the convenience-driven decline of cash acceptance creates pressure toward the mandatory cashless society that removes the option entirely.

Pay cash at the pharmacy. Pay cash at the doctor's office copay window. Pay cash for purchases you consider private. Use cash in small businesses whose owners you know and whose economic health you care about. Carry cash. Normalize cash. Its continued existence as a practical payment method depends on its continued use.

The CBDC Watchlist: Staying Alert to the Most Consequential Financial Development

President Trump's January 2025 executive order prohibiting retail CBDC development in the United States is a meaningful but not permanent protection. Executive orders can be reversed by subsequent administrations. The global CBDC infrastructure is developing regardless of American domestic policy. And the broader digital payment surveillance infrastructure operated by private companies rather than central banks continues to develop without the political scrutiny that CBDC has attracted.

Stay informed. Follow the Atlantic Council's CBDC Tracker, which monitors global CBDC development across 134 countries in real time. Understand your elected representatives' positions on digital currency and financial privacy legislation. Support the Electronic Frontier Foundation, the American Civil Liberties Union, and the organizations that are monitoring and litigating financial privacy issues at the institutional level.

When financial privacy legislation comes before your state legislature or Congress, engage. The infrastructure decisions being made now about how digital currency is architected will determine financial privacy for decades. The window to shape those decisions is open. It will not stay open indefinitely.

Part Four: The Political Fight From Citizen to Champion

Individual protective measures are necessary and you should implement them. But the structural problems described in this book the surveillance economy, the CBDC threat, the child safety crisis, the deepfake identity crisis, the cyber infrastructure vulnerability cannot be solved by individuals protecting themselves one at a time. They require political solutions. And political solutions require citizens who are organized, informed, and engaged.

This is the part of the chapter that most people skip. It is the most important part.

The Legislation That Needs Your Voice

The NO FAKES Act which would create a federal right for every American to control the use of their voice and image in AI-generated content needs to become law. It has bipartisan support. It has passed committee review. It is stalled by the ordinary inertia of a Congress that responds to constituent pressure. Contact your senators and representative in support of the NO FAKES Act. Do it now, not eventually. Congressional offices count constituent contacts. Calls are more effective than emails. Emails are more effective than nothing. Make the contact.

The Kids Online Safety Act passed the Senate 91-3 and has not yet been signed into law. It needs to complete its legislative journey. If you care about children and social media and the research is unambiguous about the scale of the harm make this call too.

COPPA 2.0 expanding child privacy protections to cover users up to sixteen and prohibiting behavioral advertising targeting minors is in the same legislative limbo. Same call, same contact.

Your state legislature matters as much as Congress in many of these areas. State privacy laws California's CCPA, Virginia's CDPA, Colorado's Privacy Act have led federal legislation by years. If your state has not passed comprehensive digital privacy legislation, find out why and find the organizations working to change it.

The Organizations Worth Supporting

The Electronic Frontier Foundation is the most important nonprofit working on digital rights, privacy, and civil liberties in the digital age. It litigates, lobbies, publishes, and advocates with a consistency and expertise that has made it a critical institutional voice for ordinary people against concentrated technological and governmental power. Supporting the EFF financially is one of the highest-leverage uses of money for anyone who cares about the issues in this book.

The Center for Democracy and Technology works specifically on child online safety, data privacy, and government surveillance issues with a policy focus. EPIC the Electronic Privacy Information Center focuses on government surveillance and corporate data practices. The Consumer Financial Protection Bureau, when adequately resourced and empowered, represents the institutional watchdog for financial data exploitation. Support it, follow it, and advocate for its independence.

At the local level: school board meetings, city council discussions about surveillance cameras and facial recognition, community conversations about smart city technologies these are the arenas where the surveillance infrastructure you live within is installed or declined. Show up. Ask questions. Demand answers about what data is collected, how it is stored, who can access it, and how long it is retained.

Vote Like Your Data Depends on It Because It Does

Every election at every level is a data privacy election, whether the candidates frame it that way or not. The officials who appoint FTC commissioners, who chair the Senate Commerce Committee, who sit on the House Energy and Commerce Committee, who lead state attorneys general offices these are the decision-makers who determine whether the surveillance economy faces accountability or not.

Research candidates' records on technology policy, not just their general party affiliation. Some of the most effective advocates for digital privacy and child online safety are Republicans. Some of the most industry-captured legislators are Democrats. The issue cuts across conventional political alignments in ways that reward voters who look at actual records rather than party labels.

Make technology policy a voting issue. Tell candidates it is a voting issue. The reason that financial industry and technology industry lobbying consistently prevails over public interest in legislative outcomes is not that the public does not care. It is that the public has not consistently communicated to elected officials that it will hold them accountable at the ballot box for failing to act. Change that communication and the calculus changes.

Part Five: The Irreplaceable Human Skills That Outlast Every Algorithm

We have spent this book examining what technology is doing to human beings. This final section asks a different question: what are human beings capable of that technology cannot touch?

The answer is more than you might expect. And developing these capabilities is not just a career strategy or a personal development project. It is the deepest and most fundamental form of taking back control reclaiming the specifically human qualities that make your existence something more than a data set to be harvested and a behavior to be predicted.

MIT's Sloan School of Management published research in March 2025 that identified what it calls the EPOCH framework five categories of uniquely human capability that AI systems cannot replicate: Empathy, Presence, Opinion and Judgment, Creativity, and Hope. The researchers were explicit that they deliberately avoid calling these soft skills. A hard skill, like solving a math problem, is comparatively easy to teach and therefore to automate. Teaching a person genuine empathy, genuine creativity, genuine hope and the specific wisdom that comes from a life lived in a human body, in human relationships, in a specific community and culture is not just harder. It may be impossible for any system that does not actually live that life.

These are your advantages. Develop them with the same seriousness that you would develop any professional skill, because in the coming decade they will be exactly that.

Empathy: The Skill That Scales

Empathy the genuine ability to understand and share the feelings of another person, to see the world from inside their experience is the irreducible foundation of every relationship, every negotiation, every act of leadership, every therapeutic conversation, every effective sale, every meaningful piece of creative work. It is the quality that makes a doctor more than a diagnostic algorithm, a teacher more than a curriculum delivery system, a leader more than a management process.

AI systems can simulate empathic responses. They can produce text that reads as empathic, that uses the right words, that follows the appropriate emotional logic. What they cannot do is actually feel cannot have the lived experience of loss or joy or fear or love that makes empathic understanding something other than a statistical pattern. This distinction matters enormously in contexts where the person receiving the empathy needs to feel genuinely seen, not efficiently processed.

Develop your empathy deliberately. Practice listening without planning your response the discipline of genuine presence in a conversation, fully attending to what the other person is experiencing before considering what you will say. Seek out people whose lives are different from yours. Read literature not self-help, not business books, but fiction that places you inside the consciousness of people whose experiences you would never otherwise access. The reader who has lived inside Anna Karenina and Atticus Finch and Celie from The Color Purple has developed a form of emotional intelligence that no AI training corpus can replicate, because it is not the knowledge of these characters but the experience of inhabiting them.

> **MIT SLOAN, MARCH 2025:** Research identified the EPOCH framework Empathy, Presence, Opinion/Judgment, Creativity, and Hope as the five uniquely human capabilities AI cannot replicate. Researchers explicitly rejected the term 'soft skills': 'It is much harder to teach a person empathy or creativity than to teach them to solve a math problem.'

Presence: The Rarest Commodity

In a world of infinite digital distraction, genuine human presence the quality of actually being fully here, in this conversation, attending to this person, available to this moment has become extraordinarily rare and correspondingly valuable.

The person who can put down the phone and be fully present in a conversation with their child, their colleague, their client, their patient is providing something that no AI assistant and no algorithmic platform can replicate. The quality of attention that real presence represents undivided, unfiltered, genuinely engaged is one of the most profound gifts one person can give another, and its scarcity is increasing as the attention economy fragments every waking moment into competing demands.

Presence is developed through practice. Not through meditation retreats or expensive programs, though those can help. Through the daily discipline of choosing to be here putting the phone in another room, maintaining eye contact, asking questions that demonstrate you were actually listening, resisting the reflex to check the notification, finishing one conversation before beginning another.

In professional terms, the person who is genuinely present in a meeting, a negotiation, a client conversation will consistently outperform the person whose attention is divided. In personal terms, the parent who is genuinely present with their children is building a relationship that no amount of screen time and digital entertainment can substitute. Presence is a skill. It is learnable. And in the age of artificial distraction, it is a superpower.

The Skilled Trades: Hands That Build What Minds Cannot Replace

There is a category of human capability that is often overlooked in discussions of the AI future because it has been undervalued in the knowledge economy of the past thirty years: skilled physical work. The electrician, the plumber, the carpenter, the mechanic, the welder, the chef, the farmer, the nurse who takes a blood pressure reading and notices something in the patient's eyes that no diagnostic algorithm would flag these are people whose work is rooted in physical reality, in the specific and unpredictable circumstances of the actual world, in a way that cannot be fully automated.

The World Economic Forum's Future of Jobs Report 2025 projects that roles requiring physical dexterity, environmental unpredictability, and hands-on problem solving are among the most resilient to automation. An electrician troubleshooting a circuit problem in a century-old house encounters conditions that no training dataset fully anticipates. A plumber working in a building where three generations of previous plumbers each made different decisions is navigating a complexity that requires judgment, adaptability, and physical skill in combination. A chef who understands the specific texture and flavor of today's produce at today's temperature in today's kitchen is doing something that AI-guided kitchen automation cannot replicate.

Skilled trades are not consolation prizes for people who could not make it in the knowledge economy. They are foundational human capabilities deeply satisfying, economically resilient, immune to offshoring, and increasingly lucrative as the supply of skilled tradespeople has declined while demand has grown. The median electrician in the United States earns more than the median holder of a four-year college degree, with no student debt and with skills that appreciate rather than depreciate as experience accumulates.

Whether your career is in the trades or not, developing practical physical competencies the ability to grow food, to repair your own home and vehicles, to build things with your hands, to produce something tangible from materials provides a form of resilience and autonomy that the subscription economy is designed to make you forget you need.

Critical Thinking and Judgment: The Skill That AI Makes More Valuable

Here is a counterintuitive truth about the age of AI: the flood of AI-generated content the essays, the reports, the analyses, the recommendations makes the ability to evaluate that content, to identify its errors and limitations, to ask the questions that expose its assumptions, more valuable, not less.

AI systems produce confident-sounding output that is sometimes wrong, sometimes misleading, sometimes subtly biased by the data they were trained on. The professional who can read an AI-generated analysis and ask: what was this trained on? What assumptions does it embed? Where is the uncertainty? What is it not accounting for? that professional provides something that no AI can provide about itself. The ability to see clearly, to reason from first principles rather than from pattern recognition, to distinguish genuine insight from plausible-sounding error, is the irreplaceable complement to AI capability.

Develop this by practicing the discipline of examining your own assumptions. By reading primary sources rather than summaries of summaries. By engaging seriously with positions you disagree with not to be convinced by them, but to understand their actual strongest form. By asking, about any claim that matters: how do we know this? What would make it false? Who benefits from believing it?

These are ancient intellectual virtues. They are the foundation of good science, good law, good journalism, good governance, good medicine. They are not automated. They may not be automatable, because they depend on a relationship to truth and to reality that is grounded in values in caring about what is actually true rather than what is computationally optimal that AI systems do not possess.

Community and Relationships: The Network That Algorithms Cannot Build

The atomization of modern life the replacement of community with platform, of neighbor with follower, of genuine mutual aid with algorithmic content delivery is one of the least discussed but most consequential dimensions of the surveillance economy's impact on human society.

Real community the kind where people know each other's names, where a neighbor notices when you have not been seen in a few days, where skills and resources and food and care circulate based on relationship rather than transaction is resilient in a way that digitally mediated social networks are not. It is also, increasingly, the technology that surveillance capitalism cannot easily commodify.

Know your neighbors. Not as Facebook contacts or NextDoor users, but as people whose faces you know, whose situations you understand, with whom you have exchanged genuine assistance. Build the kind of relationships that function during a power outage, during a medical emergency, during the kinds of infrastructure disruptions that Volt Typhoon has been positioned to execute. The communities that survive crises are the communities where people already know and trust each other before the crisis begins.

Invest in local institutions the library, the food bank, the community garden, the local business, the school that create the physical spaces where genuine community forms. These institutions are not nostalgic relics. They are the infrastructure of human resilience, and their decline in the face of digital convenience has costs that will become visible precisely when digital infrastructure fails.

Creativity: The Spark That Has No Off Switch

Human creativity is not the ability to generate novel combinations of existing patterns which is what AI does, extraordinarily well, at extraordinary scale. It is the ability to generate meaning. To make something that matters. To express something true about human experience in a form that other humans recognize and respond to. To imagine a future that does not yet exist and inspire others to build it.

Every great artist, every great scientist, every great political leader, every person who has ever changed something significant about how human beings live has done so through an act of imagination that was rooted in specific human experience: the particular suffering that drove them to find a cure, the specific injustice that demanded a response, the precise sensory memory that a painting or a song sought to preserve, the exact moment when a possibility became visible that had not been visible before.

AI can help artists make art. It can help scientists explore data. It can help leaders communicate more effectively. It cannot replace the human experience that gives creative work its meaning and its power. The creativity that matters is not the generation of novel patterns. It is the courage to make something that reveals what you actually see, what you actually feel, what you actually believe about what it means to be alive. No training corpus contains that. No algorithm produces it. It is yours, entirely, if you choose to develop and deploy it.

Protect your creative life. Make things. Write things, even if no one reads them. Build things, even if they are imperfect. Cook, garden, play music, tell stories, draw badly, sing in the car. Not because these activities will make you economically resilient though they might but because they are the expression of your specific, irreducible humanity, and that expression is the most important thing you can protect.

The Integrated Life: Putting It All Together

Everything in this chapter the tools, the financial strategies, the political engagement, the human skills can feel overwhelming when listed in full. The purpose of this final section is to reduce the overwhelm to something actionable: a starting point, a progression, a framework for building protection and capability over time without requiring that you do everything at once.

Week One: The Foundation

Download Signal and ask three important people in your life to use it with you. Set up a password manager Bitwarden is free and takes thirty minutes and begin transferring your most important accounts. Enable two-factor authentication on your email and banking accounts. Establish a family code word and communicate it to every relevant family member. Place a credit freeze with all three bureaus.

That is the first week. Five actions. Each takes between ten and sixty minutes. Together they address the most acute vulnerabilities communication interception, credential theft, financial identity fraud that affect the largest number of ordinary people. Do these five things and you are meaningfully safer than you were before.

Month One: Building the Structure

Switch your default browser to Brave and your default search to DuckDuckGo. Install uBlock Origin. Set up ProtonMail for sensitive email communications. Sign up for Mullvad or ProtonVPN. Review and audit the app permissions on your smartphone revoke location, microphone, and camera access from any app that does not genuinely need it. Review your social media privacy settings and tighten them. Request your free credit reports and review them carefully.

Quarter One: Going Deeper

Subscribe to DeleteMe or work through manual data broker opt-outs. Create a family financial protocol a documented set of rules about verification requirements for unusual financial requests, shared with every relevant family member. Review your retirement savings and ensure you are contributing at the maximum affordable level. If you carry revolving credit card debt, create an aggressive payoff plan. Research your state's privacy laws and understand your rights.

Ongoing: The Political Work

Contact your senators and representative about the NO FAKES Act, KOSA, and COPPA 2.0. Find the local organizations working on technology policy and child online safety in your community. Attend one school board meeting to understand what technology decisions are being made about your children's schools. Support the EFF. Vote in every election, at every level, having researched candidates' positions on digital rights and child online safety.

> **THE WEEK ONE CHECKLIST:** 1. Download Signal. 2. Set up Bitwarden. 3. Enable two-factor authentication on email and banking. 4. Establish a family code word. 5. Place a credit freeze at all three bureaus. Do these five things this week. Nothing else on this list matters as much as starting.

The Letter to the Person You Are Becoming

We began this book with a morning. An ordinary morning coffee, a phone check, a commute in which you were harvested dozens of times before nine o'clock by systems you could not see. We have spent six chapters examining those systems with the lights on, which is uncomfortable, which is necessary, and which is not, ultimately, the point.

The point is what you do with the light.

You are not the person this surveillance economy has decided you are. You are not the profile in Acxiom's database, the click pattern in Facebook's behavioral graph, the voice sample in an AI model's training data. You are a human being with specific experiences, specific relationships, specific skills, specific values, specific creativity a particular, irreducible, unreplicable individual whose existence generates meaning that no algorithm will ever adequately capture.

They have been trying to reduce you to data. The answer is to refuse reduction.

That refusal is not a single dramatic act. It is a thousand small, consistent choices: the app permission you revoke, the phone call you make to your elected representative, the code word you establish with your family, the book you read with sustained attention, the neighbor you actually meet, the skill you develop with your hands, the creative work you make even though no platform will reward you for it with likes.

It is the choice to be present in your actual life rather than performing in your digital one. The choice to build things that are yours rather than renting access to things that belong to platforms. The choice to know things deeply rather than skimming the algorithmically optimized surface of everything. The choice to engage politically rather than accepting the comfortable fiction that someone else is handling it.

Every previous generation faced the question of what to do with the power that threatened them. The generation that faced the industrial revolution organized labor movements. The generation that faced fascism organized armies. The generation that faced segregation organized freedom rides. They were ordinary people, most of them not heroes from birth, not uniquely courageous, but people who understood what was at stake and chose not to look away.

The threat of this generation is quieter. It does not march in jackboots or demand that you sign away your rights in a single dramatic moment. It arrives in increments in a terms of service agreement, an app permission, a convenience feature, a subscription that replaces a purchase, a platform that replaces a community, an algorithm that replaces a teacher, a deepfake that replaces a voice you trusted. It is designed to be incremental. Incremental is harder to resist than dramatic.

But resistance is possible. It is happening. The 91-3 Senate vote on child online safety. The French phone ban. The Australian social media age law. The parent movements organizing around phone-free schools. The citizens who understand CBDC and are demanding financial privacy protections. The workers who went on strike for 118 days because they refused to let their faces and voices be owned forever for one day's pay.

These are your people. This is your fight. And it is winnable not because the technology will reverse itself, not because the powerful will voluntarily surrender their advantages, but because human beings, when they understand what is at stake and act together with sustained purpose, have consistently proven capable of reshaping the conditions of their own lives.

That capacity is the most important thing that no AI will ever take from you.

Use it.

END OF CHAPTER SIX

CHAPTER SEVEN

The Fine Print and the Ant Colony:

How to Read the Contracts That Run Your Life, Why Your Wallet Is the Most Powerful Weapon You Own, and How Ordinary People Win

---

"The ants pick up the grain. We are a lot stronger than you say we are. And you know it, don't you?" Flik, A Bug's Life, 1998

A Dog Commercial That Changed the Conversation

It was February 9, 2026. Super Bowl Sunday. One hundred and twenty million Americans sat in front of their televisions for the most watched broadcast of the year, and between the touchdowns and the halftime show and the celebrity cameos, they watched a thirty-second advertisement from Ring the Amazon-owned home security company that was supposed to make them feel warm and fuzzy about technology.

The ad showed a man's beloved yellow Labrador, Milo, gone missing. His distraught face. His children's worried eyes. Then the solution: Ring's new Search Party feature. One photo uploaded to the app. A single Ring camera activates above a garage then, in a birds-eye aerial view, every camera in the neighborhood lights up simultaneously, a grid of glowing surveillance nodes, all searching, all scanning, all connected. A white targeting box appears around a dog sniffing down a street. The box turns green. Milo is found.

The ad was designed to be heartwarming. It produced the opposite reaction.

Within hours, the commercial was going viral for entirely the wrong reasons. A TikTok calling it terrifying accumulated three million views. Comments flooded social media: 'Nice way to start a mass surveillance product and label it as dog rescue.' 'AI video surveillance of your neighborhood. Constantly, every day, always watching. You can't hide or escape.' Senator Ed Markey of Massachusetts posted simply on X: 'This definitely isn't about dogs it's about mass surveillance.'

A Seattle-based artist named Maggie Butler posted a video of herself pulling her Ring camera off her porch, flipping it a middle finger, and explaining to her followers exactly why. Videos of people removing their Ring cameras went viral. One man's response tweet was quoted by hundreds of thousands: 'Government: how can we get Americans to accept constant surveillance? Americans: PUPPIES!!!!!!'

The WeRateDogs account nearly twenty million followers came out swinging: 'We all love a good, wholesome dog commercial. This is not that. This is a company putting a cuddly face on a rather dystopian reality: widespread networked surveillance by a company that has cozy relationships with law enforcement.'

Four days later, Ring quietly announced it was ending its partnership with Flock Safety the controversial police surveillance company whose license plate reader technology had been integrated with Ring's law enforcement access portal. Ring cited 'time and resources.' Nobody believed them. The implication of the timing was unmistakable: the public had pushed back, and the company had blinked.

This is the most important sentence in the opening of this chapter: the public had pushed back, and the company had blinked.

Keep that sentence. We are going to return to it.

The Dog That Revealed the Network

To understand why the Ring Super Bowl ad produced such immediate, visceral public reaction, you need to understand what it actually showed because the ad's creators, in their enthusiasm to demonstrate the feature's power, showed far more than they intended.

The aerial shot of an entire neighborhood's camera network simultaneously activating was not a dramatization for advertising effect. It was an accurate visual representation of what Search Party actually does. Ring's network, as of early 2026, encompasses cameras installed in approximately 27% of American households. In dense suburban neighborhoods the kind shown in the ad that penetration can mean that virtually every block has multiple Ring cameras covering every angle of approach.

When Search Party activates, it does not just alert nearby cameras. It coordinates them. The AI system that Ring has built trained on tens of thousands of videos to recognize dog breeds, sizes, fur patterns, and body features scans the feeds of every opted-in camera in the search radius. It applies object recognition across a distributed network of privately owned cameras that together function, precisely as the ad showed, as a comprehensive neighborhood surveillance grid.

> **THE SCALE:** An estimated 27% of American households have doorbell cameras. In typical suburban neighborhoods shown in the Ring ad, this creates overlapping coverage of virtually every public space. The Search Party feature coordinates these cameras into a unified AI-scanning network. Ring confirmed: facial recognition technology is already built into its devices.

The privacy expert who best articulated the core problem was Chris Gilliard, author of the forthcoming book Luxury Surveillance, who told 404 Media that Amazon was putting 'a cuddly face on a rather dystopian reality.' The cuddly face was the dog. The dystopian reality was this: the same AI system that identifies a yellow Labrador by its fur pattern and body shape is architecturally identical to a system that could identify a human being by their face and gait. The difference between finding a missing dog and tracking a specific person through a neighborhood is not a hardware difference. It is a software configuration.

Ring confirmed this directly. A Ring spokesperson acknowledged that facial recognition technology is already built into its devices it is used for the familiar faces feature, which sends personalized alerts when the camera recognizes specific people. The company stated this requires user permission. What it did not fully address is the question that privacy advocates immediately raised: once the infrastructure exists the cameras, the AI, the networked coordination capability, the law enforcement access portal the protections that currently limit its use to dogs and familiar faces are policy choices, not technical constraints. Policy choices can change. Technical infrastructure, once deployed, tends to remain.

The Flock Safety Connection and What It Revealed

The Ring-Flock Safety partnership that Ring quietly cancelled four days after the Super Bowl backlash deserves examination, because it is the clearest illustration of the pipeline from consumer surveillance to law enforcement surveillance that the ad's critics were warning about.

Flock Safety is a company that operates automated license plate reader cameras across thousands of American cities and towns. Its cameras, installed on streetlights and utility poles, photograph every vehicle that passes, capture the license plate, and store the data in a searchable database accessible to law enforcement agencies. As of 2025, Flock's network was operating in over 5,000 jurisdictions across all 50 states.

Investigations into Flock's data obtained through public records requests representing over 12 million searches from more than 3,900 agencies revealed a pattern of systemic abuse. Police departments used Flock cameras to track protesters at political demonstrations. Flock data was used to target Romani people with discriminatory stops. Immigration enforcement agencies used the system for nationwide vehicle tracking.

The planned Ring-Flock integration would have linked Amazon's massive consumer camera network covering the front doors and driveways of tens of millions of American homes with Flock's law enforcement surveillance infrastructure. The result would have been a combined system of almost incomprehensible reach: Ring's cameras covering private residential property feeding into a system already used by ICE, CBP, and thousands of local police departments for nationwide tracking.

Ring said the integration never launched. Ring ended the partnership under consumer pressure. But the partnership was planned, announced, and only cancelled when public reaction made its continuation politically untenable. The intention was documented. The infrastructure was partially built. The cancellation was not a change of values it was a business decision driven by customer backlash.

That distinction matters enormously, because it tells you exactly how the system changes: not from the inside, not from a spontaneous corporate conversion to privacy values, but from the outside, through the organized, sustained, economically consequential pressure of customers who understand what is being built and refuse to pay for it.

The Default Setting That Built an Empire

One of the most quietly significant details in the Ring Search Party story is the question of default participation. According to reporting from the Biometric Update and multiple privacy advocates, Search Party was enabled by default on certain compatible Ring devices. Users had to actively opt out if they did not want their cameras participating in the neighborhood scanning network.

This is the default effect, which we first encountered in Chapter One's discussion of dark patterns. When a setting is on by default, the overwhelming majority of users will never change it not because they affirmatively want the setting, but because changing defaults requires awareness, effort, and time that most people do not bring to their relationship with consumer technology.

When a surveillance network is opt-out rather than opt-in, the network's effective size is determined not by how many people actively choose to join it but by how many people fail to notice they are already part of it. Ring did not need everyone to enthusiastically consent to Search Party. It needed most people to never check their settings. And most people never check their settings.

This is why the fine print matters. This is why the default settings matter. This is why the terms of service that you have been skipping for your entire digital life contain information that is, increasingly, a matter of your physical safety and privacy in the real world, not just your abstract digital profile.

The Contracts That Run Your Life: A Field Guide to Reading What You Sign

You have signed hundreds of them. Maybe thousands. The terms of service agreements, privacy policies, end user license agreements, and terms and conditions documents that gate access to every digital service, every app, every device, every online account. You have signed them by clicking a button that said Agree, in approximately one second, after scrolling past paragraphs of dense legal text at a rate of comprehension that no human being could achieve at any speed.

They are not meaningless. They are, in fact, the most consequential legal documents that most ordinary people sign in their lives more consequential, in aggregate, than any lease or employment contract, because they govern the full scope of how the companies that mediate your digital life treat your data, your likeness, your privacy, your communications, and your legal rights.

And they are specifically designed to be unread.

This section does not tell you to read every terms of service document in full. That was established as impossible in Chapter One seventy-six work days per year, for the average American. What it tells you is what to look for, what questions to ask, and which specific phrases and provisions should trigger immediate alarm regardless of which company is deploying them.

The Five Phrases That Should Stop You Cold

You do not need to read every word of a terms of service document to understand its most important provisions. You need to learn to search for five categories of language that represent the highest-risk grants of permission that companies routinely bury in legal text.

The first is perpetual license. When a terms of service grants the company a 'perpetual, irrevocable, worldwide, royalty-free license' to use your content, your likeness, or your data, it is claiming the right to use what you have provided, forever, for any purpose, without additional compensation, regardless of whether you later delete your account. This language appears in the terms of service of virtually every major social media platform. It means that the photographs you posted, the videos you shared, the voice recordings you uploaded, remain available for the company's use even after you believe you have withdrawn your consent by deleting your account. As we established in Chapter Three, this is the mechanism that makes your social media archive a permanent resource for AI training, identity replication, and commercial use.

The second is we may share with third parties. This phrase, and its countless variations affiliates, partners, service providers, business successors is the legal mechanism for the data broker ecosystem described in Chapter One. When a privacy policy tells you that your data may be shared with unspecified third parties for unspecified business purposes, it is telling you that your data will be sold. The vagueness of 'third parties' and 'business purposes' is deliberate. It grants permission for data sharing that, if described specifically, would horrify most users.

The third is we may use your data to improve our services. This innocuous phrase is the standard language through which companies grant themselves permission to use your data your messages, your photos, your behavioral patterns, your voice recordings to train AI models. Every time you have interacted with a digital assistant, a photo recognition system, a recommendation algorithm, or a voice recognition service, you have almost certainly provided training data under this clause. The 'improvement of services' is not a consumer benefit. It is the construction of commercial AI systems of substantial value, built from your data, without your specific knowledge or additional compensation.

The fourth is we may update these terms at any time. This clause, combined with a provision that your continued use of the service constitutes acceptance of the updated terms, means that any privacy commitment the company makes today can be unilaterally revoked tomorrow. The terms you agreed to when you signed up are not the terms that govern your relationship with the company today. They have almost certainly been updated, often significantly, and your acceptance was your continued use of the app.

The fifth, and most immediately actionable, is you agree to arbitration. This clause, buried in the dispute resolution section of most major consumer technology terms of service, requires you to resolve any dispute with the company through private arbitration rather than through the courts. It typically prohibits class action lawsuits, meaning you cannot join with other affected consumers to bring a collective claim. It means that if the company violates your privacy, steals your data, or causes you harm through its practices, your recourse is a private proceeding before an arbitrator chosen from a panel in which the company has systematic advantages. It is, functionally, a waiver of your right to hold the company publicly accountable in a court of law.

> **SEARCH FOR THESE:** Perpetual, irrevocable license | Share with third parties | Improve our services | Update terms at any time | Mandatory arbitration. These five phrases represent the highest-risk grants of permission in consumer technology contracts. Find them. Understand what they mean. Then decide whether the service is worth what it actually costs.

The Settings Audit: What Is Already On

Right now, on the devices you own and the accounts you hold, there are settings enabled that you did not affirmatively choose settings that were activated by default when you first set up the service, or that were turned on through a terms of service update you did not read, or that were presented to you through a dark pattern that steered you toward consent you did not intend to give.

The settings audit is not a one-time event. It is a periodic practice conducted every six to twelve months, or whenever a company makes a significant product change that prompts a notification you would ordinarily dismiss.

For your smartphone: go to Settings, then Privacy, then each major category Location Services, Microphone, Camera, Contacts, Photos, Health. For each app listed under each category, ask: does this app actually need this permission to function? A flashlight app does not need location access. A recipe app does not need microphone access. A game does not need your contacts. Revoke every permission that is not genuinely necessary. The apps that stop functioning when you revoke an unnecessary permission have told you something important about why they wanted that permission.

For your smart home devices speakers, doorbells, cameras, televisions go through the companion apps and disable data sharing options, voice recording history, personalized advertising, and any surveillance or sharing features that were enabled by default. Amazon Alexa, Google Nest, Ring, and most smart TV platforms have privacy dashboards that allow you to review and limit data collection. They are not prominently advertised. They exist because regulations in various jurisdictions required them. Find them.

For your social media accounts: review the advertising preferences settings, which reveal what categories of behavioral data the platform has collected about you and is using to target advertisements. On Facebook, this section is called Ad Preferences. On Google, it is My Ad Center. What you find will almost certainly surprise you. The specificity of the behavioral profile these dashboards reveal interests, life events, income bracket inferences, political affiliation signals is one of the most direct demonstrations available of what the surveillance economy actually knows about you.

The New Account Checklist: Before You Click Agree

For any new service, app, or device you consider adopting, apply a brief but non-negotiable checklist before you click agree.

Where is this company headquartered, and what jurisdiction's privacy laws apply? A company headquartered in the European Union operates under GDPR, with meaningful enforcement. A company headquartered in a U.S. state without comprehensive privacy law operates under far weaker protections. A company headquartered in a country with minimal privacy law and potential government access to corporate data China, in particular operates under an entirely different risk profile.

Does this service require the data it is asking for? Does a fitness tracker need your email contacts? Does a photo editing app need your location? Does a news reader need microphone access? If the data the app requests is not obviously necessary for the service it provides, assume it is being collected for a secondary purpose behavioral profiling, advertising, data brokering that you are not being told about.

Is there a privacy policy that is comprehensible and specific? A privacy policy that uses vague language about 'business purposes,' 'partners,' and 'service improvement' without specificity is a privacy policy designed to permit everything while appearing to limit something. Look for policies that specifically identify what data is collected, how long it is retained, who it is shared with, and what your rights are to access and delete it.

Can you use the service without creating a full account? Many apps and services can be used without the full data collection that account creation enables. Using a service as a guest or using a privacy-focused email alias service like SimpleLogin to create disposable email addresses for account registration reduces your exposure to the full behavioral profile that linked, persistent accounts enable.

The Subscription Car: Chapter Four Comes to Your Driveway

In Chapter Four, we examined the subscription economy's systematic replacement of ownership with access the way that software, media, and services that you once bought outright have been converted into recurring revenue streams that extract continuous payments for what used to be a single transaction.

The car sitting in your driveway is the subscription economy's most audacious frontier.

In 2022, BMW made the news all of it bad news for offering heated seats as a subscription feature. The car was delivered with the heated seat hardware already installed, physically present in the vehicle you owned. But the software that activated that hardware was locked behind a monthly fee. You had paid for the seat. You had not paid for the right to warm the seat that you had already paid for.

The public reaction was swift, ferocious, and unified across every political and demographic line. The fundamental indignity of paying for hardware you owned but being charged to use it struck virtually everyone as an obvious and intolerable absurdity. BMW faced weeks of mockery, criticism, and most importantly defection. Customers and prospective customers announced publicly that they would not buy a vehicle that subjected them to this model. The company quietly reversed course on hardware-based subscriptions, with executives later admitting, with characteristic understatement, that heated seats were 'probably not the best way to start' with subscription features.

But BMW did not abandon subscriptions. It retreated from the most visible and obviously offensive application while continuing to push the model forward. And every other major automaker watched BMW's experience, drew the lesson don't start with heated seats and continued developing their own subscription programs.

THE AUTOMOTIVE SUBSCRIPTION MARKET: The in-car subscription services market was valued at $9.2 billion in 2024 and is projected to reach $791 billion by 2032. This growth trajectory represents automakers' intention to convert your vehicle from a product you own into a service you access continuously, for as long as you drive.

What Is Currently Being Locked and Charged For

The scope of automotive features that are currently, or have recently been, offered as subscriptions is wider than most car buyers know. Understanding the landscape is the first step to not consenting to it.

BMW, having abandoned the heated seat subscription under pressure, has continued offering subscriptions for adaptive cruise control assistance features, adaptive headlights in European markets, and an annual subscription for unlocked performance on certain electric models. The company has been explicit that software-based subscriptions are the future of its revenue model.

Volkswagen has offered power upgrades by subscription on its ID.3 electric vehicle in UK markets the car is manufactured with the hardware to produce 225 horsepower, but the base model is software-limited to 198 horsepower. Paying a monthly subscription unlocks the power that the car was built to produce. You are not paying for additional capability. You are paying for permission to use the capability you already own.

Mercedes-Benz has offered a subscription for enhanced rear-wheel steering on its EQS the car contains the hardware for a 10-degree steering angle, but owners in some markets pay to unlock the full range from the factory-limited 4.5 degrees. Mercedes also offers acceleration upgrades by annual subscription of $600 to $900 on its EQ electric models paying, in effect, for permission to access the performance that the motor you purchased is capable of producing.

Tesla's Full Self-Driving package is available as a $99 monthly subscription. Tesla has moved away from offering the one-time purchase option, converting what was a large optional expenditure into a permanent recurring revenue stream. The vehicle you purchased contains the hardware that FSD uses. The software, updated continuously via over-the-air updates, is rented.

Mazda requires a $10 monthly subscription for connected services including remote start the physical hardware that starts your car from a distance is built into the vehicle. The digital key that activates it costs ten dollars per month, every month, for as long as you want to use a feature you purchased when you bought the car. When Mazda customers pushed back, some of them went further than complaints: in 2024, Mazda owners hacked into their own vehicles with software workarounds to access features without paying the monthly fee. They were accessing hardware they owned, in a car they owned, to use a feature they had paid for. The company called it a violation of terms of service.

> **THE PATTERN:** BMW, VW, Mercedes, Tesla, Mazda. Every major automaker is building software locks into physical hardware you own, then charging monthly for the key. The automotive subscription market is projected to grow 86-fold by 2032. This is not a niche experiment. It is the industry's intended future for everyone who buys a car.

The Data Your Car Is Collecting

The subscription model is not the only way your new car is being monetized after the point of sale. Modern connected vehicles collect data enormous quantities of it and transmit it to manufacturers and third parties in ways that most car buyers have never been told about and would not consent to if they were.

Your connected car knows your precise location at all times. It knows when you started the engine, where you drove, how long you were there, and when you returned. It knows how fast you drove, whether you used seat belts, how aggressively you braked and accelerated. It knows whether multiple people regularly ride in your car. Through cabin microphones designed for voice assistants and hands-free calling, it may record ambient audio. Through in-cabin cameras designed for driver monitoring systems, it may record video of you and your passengers.

This data is transmitted to the manufacturer. What happens to it from there varies by company and jurisdiction, but in the United States where there is no comprehensive automotive data privacy law it is sold to data brokers, insurance companies, and other commercial entities with minimal restriction. Insurance companies have been documented using data purchased from connected car manufacturers to retroactively reprice policies, cancel coverage, or deny claims based on driving behavior that customers did not know was being recorded.

General Motors was found in 2024 to have sold detailed driving behavior data from millions of customers to insurance companies without clear disclosure data that was collected through its OnStar connected services system. The data included specific instances of hard braking, rapid acceleration, and other behaviors that insurers used as grounds for premium increases. GM customers had not been told, in any prominent or clear way, that subscribing to OnStar meant consenting to the sale of their driving behavior to insurance companies.

Your car, in the connected vehicle age, is a data collection device that happens to have an engine. The subscription fees are the visible part of its ongoing monetization. The data harvest is the invisible part. Both are happening simultaneously.

Asking the Questions Before You Buy

The consumer who walks into an automobile dealership in 2026 and asks the right questions is in a fundamentally better position than the consumer who discovers subscription features and data collection practices after purchase. The questions to ask before signing:

What features in this vehicle require a subscription after purchase? Ask for a written, comprehensive list. If the dealer cannot or will not provide one, that is information. Find the complete list on the manufacturer's website before signing anything.

Which of these subscriptions transfer to a new owner if I sell the car? Many subscription features do not transfer the new owner must start their own subscription, which effectively reduces the resale value of the vehicle. A car with significant locked subscription features is worth less in the used market than it appears.

Can I purchase any of these features as a one-time payment rather than a subscription? Some manufacturers offer this option. BMW's heated seat reversal was specifically a reversal to one-time purchase for hardware-based features. Ask explicitly for the buy-once option before assuming subscription is the only path.

What data does this vehicle collect and transmit, to whom, and for what purpose? Ask for the connected services privacy policy in writing. Read the sections on data sharing with third parties. If the answer is that driving behavior data may be shared with partners including insurance and financial services companies, you are being told that your driving will be sold to your insurer. Decide whether you accept that before you sign.

The Ant Colony Principle: Economic Power in the Hands of the Many

In the 1998 Pixar film A Bug's Life, the grasshoppers control the ant colony through intimidation and the manufactured belief that the ants have no power. The ants vastly outnumber the grasshoppers. They vastly outproduce them. The entire system that benefits the grasshoppers depends entirely on the ants' labor and compliance.

The film's central revelation delivered by Flik, the ant who refuses to accept the framing is simple and devastating: the ants are not powerless. They are the power. The grasshoppers' entire advantage rests on the ants not knowing this.

This is not a perfect analogy for the relationship between ordinary consumers and the technology and automotive industries. But it is close enough to be useful.

There are approximately 335 million Americans. There are four major technology platforms that dominate the surveillance economy. There are perhaps a dozen major automotive manufacturers pursuing the subscription model. There are a handful of companies that own the vast majority of cloud storage, smart home devices, and connected consumer technology.

The companies are larger in aggregate resources. They have more lawyers, more lobbyists, more money for advertising. But they have a profound and fundamental dependency that their lobbying and advertising are designed to obscure: they need customers. Without customers, they have nothing. Their valuations, their revenues, their ability to fund lobbying and advertising all of it flows from the decisions that ordinary people make about what to buy and what to refuse.

We have seen this principle in action, documented and recent, throughout this book. BMW reversed its heated seat subscription under consumer pressure. Ring cancelled its Flock Safety partnership under consumer pressure. Social media companies have made privacy policy changes, product reversals, and public commitments specifically in response to organized consumer backlash. The TAKE IT DOWN Act was accelerated by public outrage. KOSA passed the Senate 91-3 in response to years of parent advocacy.

These are not anomalies. They are the normal operation of consumer and political power when it is organized, persistent, and understood by the people exercising it.

Boycott as Strategy: When Withdrawal Speaks Louder Than Arguments

The boycott the organized, collective withdrawal of consumer participation from a product, company, or practice is one of the oldest and most effective tools of nonviolent economic pressure available to ordinary people. It requires no resources beyond the willingness to not buy something. It leaves no legal exposure. It does not require access to political power. It requires only numbers.

The strategic boycott is not the same as impulsive consumer anger. Impulsive anger produces a spike of negative social media posts that a company can wait out, knowing that attention spans are short and competing stories will dominate tomorrow. The strategic boycott has specific features that make it economically consequential rather than merely emotionally expressive.

It is specific. It targets a specific product, feature, or practice rather than a company in its entirety. BMW's heated seat backlash was specific it targeted a specific feature that was perceived as obviously unjust. The specificity made it impossible for BMW to dismiss as generic anti-corporate sentiment. The company understood precisely what behavior was being sanctioned.

It is sustained. A boycott that lasts a news cycle changes nothing. A boycott that becomes a persistent consumer preference I will not buy a car with locked hardware features, I will not use a smart home device with law enforcement data sharing creates a durable market signal that product planners and revenue forecasters cannot ignore.

It is communicated. A consumer who quietly switches products without explaining why provides no signal. A consumer who explains, publicly and specifically, why they are switching gives the company and its competitors information that drives change. The Ring camera removal videos went viral not because people were removing cameras but because they were explaining why.

It is coordinated. A single consumer's withdrawal is statistically invisible. The coordination of withdrawals through organized campaigns, through social media, through consumer advocacy organizations creates the scale necessary for corporate response. BMW did not reverse its heated seat subscription because one customer complained. It reversed because the complaint was universal and sustained across markets.

> **THE RECORD:** BMW reversed hardware subscription model: consumer pressure. Ring cancelled Flock Safety integration: consumer pressure. Facebook changed privacy settings after Cambridge Analytica: regulatory and consumer pressure. Multiple platforms banned deepfakes following Taylor Swift incident: consumer pressure and legislative threat. The pattern is consistent. Organized consumer power changes corporate behavior.

The Products Worth Reconsidering

This is not a call to live off the grid or to refuse all technology. It is a call to make deliberate choices about which products and services genuinely serve your interests and which are primarily vehicles for surveillance, data extraction, or the subscription conversion of things you should own outright.

Smart home devices with persistent microphones Amazon Alexa, Google Nest Hub, and similar products place always-on audio capture devices in your most private spaces. The convenience of voice control is real. The surveillance architecture they represent is also real. The question is whether the convenience justifies the access you are granting. If you use these devices, review the voice history settings regularly, enable the privacy modes that exist, and understand that the recordings are stored on manufacturer servers where they are subject to law enforcement requests and potential data breaches.

Ring and similar law enforcement-integrated doorbell cameras deserve particular reconsideration in light of everything the Super Bowl ad revealed. There are privacy-respecting doorbell camera alternatives Eufy, Reolink, and other brands that offer local storage rather than cloud storage and do not maintain law enforcement partnership portals. The security function of a doorbell camera does not require integration with a coordinated neighborhood surveillance network and a pipeline to police departments. If you want a doorbell camera, choose one that keeps your footage under your control.

Connected vehicles with opaque data collection practices and aggressive subscription models deserve the same consumer pressure that BMW's heated seats received. Manufacturers that are transparent about data collection, that offer robust privacy controls, and that provide one-time purchase options for hardware-based features should be rewarded with purchases. Manufacturers that pursue the most aggressive extraction models should be sanctioned with the most visible, most communicated withdrawals.

Social media platforms that have demonstrated knowing harm to children while prioritizing engagement metrics deserve systematic pressure from parents not just complaints, but the organized, communicated withdrawal of both their own participation and their children's. Every parent who removes their child from a platform and explains publicly why is creating a signal that advertisers, product managers, and boards of directors can see and cannot easily dismiss.

The Companies That Deserve Your Money

The consumer power argument is not only negative not only about what to refuse. It is equally about what to support. The companies that build privacy-respecting products, that offer transparent data practices, that treat subscription models fairly, that compete for your business by providing genuine value rather than capturing you through addiction and lock-in these companies deserve your deliberate, communicated support.

Proton, the Swiss privacy technology company, has built an ecosystem of email, VPN, cloud storage, calendar, and password management that competes directly with Google's product suite while operating on a genuine privacy-respecting model. Every paying Proton subscriber is funding an alternative architecture for the digital economy. Proton's existence and growth is evidence that privacy-respecting business models can compete.

Signal is free and non-profit. Supporting it financially it accepts donations sustains the most important privacy communications infrastructure available to ordinary people.

Local businesses, local banks, local news organizations, local institutions these are the economic actors who have the least capacity to extract behavioral data at scale and the most direct accountability to the communities they serve. Every dollar spent at a local retailer rather than Amazon, every account held at a credit union rather than a megabank, every newspaper subscription to a local paper rather than a national digital platform is a small but real redirection of economic power toward accountability.

The Fine Print of Daily Life: Small Decisions That Compound

The thesis of this chapter and, in some ways, of this entire book is that the decisions that matter most are not the dramatic ones. They are the small ones, made daily, that compound over time into a posture, a culture, a set of norms that either accommodate the surveillance economy or resist it.

The surveillance economy is built on small decisions. The default setting left at its factory position. The terms of service clicked through without reading. The convenience feature accepted without considering its cost. The subscription agreed to without asking whether a purchase option exists. The smart home device installed without reviewing the privacy policy. The car bought without asking what data it collects.

Each of these decisions, individually, is small. No single decision defines your privacy posture or determines your vulnerability. But they accumulate. The person who clicks Agree on every terms of service, leaves every default setting in place, accepts every subscription model, and installs every convenience device has, over years, constructed a comprehensive infrastructure of surveillance around themselves not because they chose to be surveilled, but because they chose convenience each time the choice was presented.

The alternative is not paranoia. It is not the refusal of all technology or the rejection of all convenience. It is the habit of a brief, deliberate pause before the click, the installation, the agreement, the subscription. The habit of asking: what does this actually cost? What am I actually agreeing to? What data does this collect? What can I do with this information?

That pause cultivated consistently, applied habitually is the difference between being a passive subject of the surveillance economy and an active participant who understands the terms on which they are engaging.

The Conversation That Multiplies Your Power

Your individual choices, maintained privately, change your own situation. Your individual choices, communicated to others, change the situation for everyone.

The Ring camera removal videos went viral because people explained their reasoning. The BMW heated seat backlash spread because consumers who understood the problem articulated it clearly in spaces where others could hear them. The parent movements around phone-free schools work because parents talk to other parents. The legislative momentum around KOSA and the NO FAKES Act was built by constituents who contacted their representatives and by advocacy organizations that amplified individual concerns into political pressure.

Every conversation you have about these issues with family members, with neighbors, with colleagues, with friends is an act of civic infrastructure building. You are expanding the number of people who understand what is happening and can therefore make informed choices. You are creating the conditions for the coordinated action that makes individual choices consequential at scale.

This is not a burden. It is, if you let it be, the most natural thing in the world the human impulse to share important information with the people you care about. The grandmother who tells her friends about the family code word that protects against voice clone scams is doing something as old as community itself: warning the people around her about a danger she has learned to see.

The Terms of Service for the World You Want to Live In

There is a terms of service agreement that has not been written yet. It is the agreement between the generation alive right now and the future that will be built on the decisions they make. It does not have a click-through button. It does not have language designed by lawyers to obscure its meaning.

Its terms are simple. They are the terms that every chapter of this book has been building toward.

Your data belongs to you. Your face belongs to you. Your voice belongs to you. The hardware in the products you purchase belongs to you. The privacy of your home and your communications belongs to you. The childhood of your children belongs to them. The infrastructure that sustains the society you live in belongs to all of us, and its protection is a collective responsibility that cannot be outsourced to corporations whose interests are not aligned with ours.

These are not radical terms. They are the terms that any reasonable person, asked to describe a fair relationship between human beings and the technology they use, would derive from first principles. They are the terms that the famous and powerful have already secured for themselves through organization, lobbying, and the exercise of economic and political power.

They are the terms that ordinary people are owed. That we have not yet fully received them is not a permanent condition. It is the current state of an ongoing negotiation a negotiation in which your participation is not optional, because your absence is itself a choice, and it is a choice that benefits the parties on the other side of the table.

The Reckoning: Who Is More Afraid of Whom?

Here is something worth sitting with: the companies that profit from your data spend enormous amounts of money on lobbying, on advertising, on user experience design, and on legal teams specifically because they are concerned about what happens when enough ordinary people understand what they are doing.

They paid between eight and ten million dollars for a thirty-second Super Bowl advertisement to normalize neighborhood surveillance. That is not the behavior of an industry that is confident its practices will withstand public scrutiny. That is the behavior of an industry that understands it needs to move fast, establish infrastructure before resistance can organize, and make the surveillance feel inevitable and harmless before the reckoning arrives.

The reckoning is arriving anyway. The Ring ad backfired. The Flock partnership was cancelled. The camera removal videos went viral. Senator Markey's tweet resonated because it articulated something millions of people felt instantly and specifically when they saw the aerial shot of the neighborhood surveillance grid activating for a dog.

BMW's executives admitted they had 'misread the room' on heated seats. They were not wrong about what they wanted to do they were wrong about how fast they could do it before customers noticed and resisted. The timeline of the subscription economy's advance into automotive hardware has been slowed, not stopped, by consumer resistance. It will accelerate again as soon as resistance fades.

The technology industry's greatest fear is not regulators, who move slowly and can be lobbied. It is not journalists, who can be managed with careful PR. It is not even legislators, who respond to pressure from both sides. Its greatest fear is the informed, organized, persistent consumer who understands what is being built, communicates that understanding to others, and makes spending decisions accordingly because that consumer cannot be lobbied, managed, or out-spent. They can only be served or lost.

The Ants Have Always Won

Flik was right. The ants outnumber the grasshoppers. The colony's labor creates everything the grasshoppers consume. The moment the ants understand this the moment they stop accepting the framing that they are powerless, that the arrangement is natural and inevitable, that there is no alternative the grasshoppers' advantage evaporates.

The surveillance economy is not natural. It is not inevitable. It was built through specific decisions by specific people, at specific moments, when the conditions permitted those decisions to go unchallenged. It can be rebuilt differently through different decisions, made by different people, under different conditions that we have the ability to create.

We are more than three hundred and thirty-five million people. The companies extracting value from our data, our faces, our voices, our labor, our housing, our vehicles, and our children's futures number in the dozens. The asymmetry of resources runs one direction. The asymmetry of numbers runs the other.

The ant colony principle is not metaphor. It is arithmetic. When enough of the ants understand the arrangement and refuse its terms, the grasshoppers have no alternative. They have never had an alternative. The arrangement has always depended entirely on the ants not knowing their own power.

Now you know.

The Final Checklist: What You Do Today

Every chapter of this book has ended with action. This one is no different. But this final list is not a technical checklist. It is a commitment to yourself, to the people around you, and to the future that is still being built.

Read the next terms of service you encounter. Not all of it. But search for the five phrases. Read the data sharing section. Read the arbitration clause. Decide with information you actually have.

Audit your settings this week. Your phone. Your Ring camera if you have one. Your smart speaker. Your connected car app. Find what is on by default that you did not choose. Change what you can. Understand what you cannot change.

Ask the subscription question before your next major purchase. For your next vehicle, your next software, your next appliance, your next device: what features require an ongoing subscription? Can I buy them outright? What data does this product collect and who does it share that data with?

Have one conversation this week about something in this book. With a parent. A sibling. A neighbor. A colleague. Not a lecture a conversation. Share one thing that surprised you, alarmed you, or made you want to act differently. The conversation that multiplies your power begins with this week's one conversation.

Contact one elected representative about one issue from this book. The NO FAKES Act. Child online safety. Digital privacy legislation. Your state's privacy law or its absence. One call. One email. One constituent contact, added to the pile of constituent contacts that moves legislation and creates political accountability.

Cancel one subscription you accepted without fully understanding. Review your recurring charges. Find the one that is not genuinely serving you, that represents a conversion of something you should own into something you rent, or that belongs to a company whose practices conflict with your values. Cancel it. Tell them why.

Support one privacy-respecting alternative. Switch to Signal. Switch to Brave. Switch to DuckDuckGo. Subscribe to Proton. Donate to the EFF. Purchase from a company whose business model does not depend on your surveillance. Make the economic gesture that says: this is the kind of product I will pay for, and this is the kind of product I will not.

END OF CHAPTER SEVEN

DIGITAL HOSTAGE

Chapter One: The New Surveillance Economy You Are the Product

Chapter Two: The Government You Didn't Vote For State Surveillance, the Typhoon Hackers, and the War Already Being Fought Inside America's Walls

Chapter Three: The AI Impersonation Crisis When Your Face, Voice, and Identity Are No Longer Yours

Chapter Four: Your Money, Their Control The Cashless Society, the Ownership Crisis, and the Slow Erasure of Ordinary Wealth

Chapter Five: The Children We Are Raising What We Are Handing Them, What We Are Taking From Them, and the World We Could Still Build

Chapter Six: Taking Back Control The Complete Playbook for Protecting Yourself, Your Family, Your Wealth, and Your Humanity in the Digital Age

Chapter Seven: The Fine Print and the Ant Colony How to Read the Contracts That Run Your Life, Why Your Wallet Is the Most Powerful Weapon You Own, and How Ordinary People Win

Now you know. Use it.

Page

www.ingramcontent.com/pod-product-compliance
Lightning Source LLC
LaVergne TN
LVHW090317210926
848480LV00001B/277